BETRAYED by NATURE

BETRAYED by NATURE

THE WAR ON CANCER

ROBIN HESKETH

palgrave
macmillan

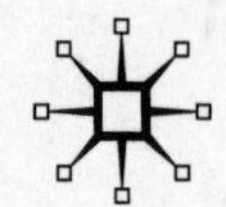

BETRAYED BY NATURE

First published in 2012 by PALGRAVE MACMILLAN® in the U.S.—a division of St. Martin's Press LLC, 175 Fifth Avenue, New York, NY 10010.

Where this book is distributed in the UK, Europe and the rest of the world, this is by Palgrave Macmillan, a division of Macmillan Publishers Limited, registered in England, company number 785998, of Houndmills, Basingstoke, Hampshire RG21 6XS.

Palgrave Macmillan is the global academic imprint of the above companies and has companies and representatives throughout the world.

Palgrave® and Macmillan® are registered trademarks in the United States, the United Kingdom, Europe and other countries.

ISBN: 978-0-230-33848-7

Library of Congress Cataloging-in-Publication Data

Hesketh, Robin.
Betrayed by nature : the war on cancer / Robin Hesketh.
p. cm.
Includes bibliographical references and index.
ISBN 978-0-230-33848-7
1. Cancer. 2. Cancer—Research. I. Title.
RC275.H47 2012
616.99'4027—dc23

2011049931

A catalogue record of the book is available from the British Library.

Design by Letra Libre, Inc.

First edition: May 2012

10 9 8 7 6 5 4 3 2 1

Printed in the United States of America.

CONTENTS

PREFACE

FOR YEARS I'VE THOUGHT OFF AND ON HOW USEFUL IT WOULD BE TO have a book that explained in the most simple way everything we think we know about cancer—certainly for me, and also for lots of people who aren't scientists but are just curious about biology or have been brought face to face with cancer and want to do battle on the most even terms they can arrange. However, for most of that time it seemed too complicated, and I mastered the art of pushing such thoughts to the dim recesses and getting on with the next experiment. Then, one day not so long ago, I was pedaling along Trumpington Road in the general direction of my lab when it dawned on me that I'd been making the classic scientist's error: I'd allowed detail to obscure basic simplicity.

That's not to deny that cancer *is* complicated—a student once wailed plaintively to me that "there are too many genes." It's worse than that, I might have replied: one of its fascinations is that just about any aspect of animal biology relates to cancer in some way and, of course, the urgency of new discoveries derives from their being right at the interface between "basic research" and clinical practice, what doctors do to earn their keep. But if you can't explain to another intelligent but non-scientific being what the basic story is in a way that they can follow and that prompts sensible questions—and that includes the medical aspects (docs take note)—how are you going to take the subject forward?

Cancers are diseases—some of them particularly unpleasant—but they are in fact nothing more than an abnormal growth of cells. We've all got them in some form: moles are just unusual clumps of cells, so they're not always dangerous. A cell is the fundamental unit of life (it can reproduce itself): it's a sac (usually called a membrane) that contains our genetic code and everything required to make a complete animal from one cell. They were christened in the seventeenth century by the all-round bright spark Robert Hooke from the

Latin for a "small room"—which is what, peering down his microscope, he thought the cells in a piece of cork looked like. Cells are full of water and other molecules—the bits and pieces that make them work. Molecules are groups of atoms (the basic unit of matter) held firmly together so that different molecules have individual features. Water (which we often call H_2O) is indeed two atoms of hydrogen stuck to one of oxygen: oxygen—the stuff we breathe—is two atoms of oxygen per molecule, so it's O_2.

Most of the molecules in cells are made by joining lots of smaller molecules together. Our genetic code comes in the form of DNA (deoxyribonucleic acid), which has three thousand million linked molecular units, the whole thing being called the genome. The "units" that are glued together to make nucleic acids are called "nucleotides" and DNA is made from only *four* different nucleotides. They differ because each nucleotide has a bit called a "base," the four that appear in DNA being denoted A, C, G, and T. You can look these up in the glossary, but the only thing that really matters is that the sequence of these "bases" carries the genetic code. The vast length of DNA is split up into shorter stretches called "chromosomes." Within chromosomes sit genes—regions of DNA where the four-letter code directs the order in which the units that make proteins are stuck together. Proteins are the machines that do all the work in cells and make living things what they are.

These then are the fundamental parts of life, and we need to be familiar with them to be able to think about how we work. We can then begin in Chapter 1 with a little history of cancer before looking at patterns of different types of cancer around the world and what they suggest about causes. After that we'll take up the molecular theme again by considering how "DNA makes proteins." This highlights the importance of the four-letter sequence of the "genetic code" and what it is that makes proteins such fantastic structures that they are indeed the "machines of life." We'll also explain the most astonishing fact of life that we merely mentioned just now—that, from the information encoded by just four bases, A, C, G, and T, all animals and plants are made.

From genes and what they make we then turn to cells—what they are and how being able to grow them in the laboratory has made such a big contribution to medicine. One of the many privileges and pleasures of a life in science is following in Hooke's footsteps and looking down a microscope to see "how my cells are getting on"—always a beautiful sight. However, one of the clever things about cells that scientists have yet to explain is how, when a few of them come together and make a human being, they may still be beautiful (OK, not always) but they also frequently become weird, eccentric, or even downright barmy. Perhaps it's not surprising that in many ways cancers resemble the bodies they colonize: some are pretty harmless (we even call them "benign"); others are distinctly malevolent and incredibly smart in the way they achieve their ends. Much remains to be discovered about how cancers establish themselves but, even so, the last thirty years have revealed a great deal, particularly about genes and what goes wrong with them to make a normal cell abnormal. Turning to that subject leads us into what it is about cancer cells that makes them potentially lethal and, in particular, how tumor cells manage to spread around the body.

In the last two chapters we focus mostly on why unraveling the secrets of DNA is so vital, beginning by briefly telling the story of how human DNA was first sequenced—an achievement that has ushered in the greatest revolution in the history of medicine. In the last chapter we look at where we stand on the detection and therapy fronts, how drug treatments have evolved and how the remarkable era we're entering is already transforming the way we deal with cancer.

So while this story is partly for folk who have been struck by cancer or indeed other diseases, either themselves or in their loved ones, in the hope that a little understanding will help, it's also for everyone else, scientists (be gentle with me) and non-scientists.

The plan is to tell the story about cancer with a light touch, much helped, I trust, by saying a little about the people who have contributed and who continue writing it. I pointed out just now that, contrary to the widely portrayed view, scientists are human and sometimes they combine being quite clever with odd outbreaks of pottiness, all of which makes life in a lab distinctly entertaining. As my first witness I call John Burdon Sanderson Haldane, a predecessor of mine in the Department of Biochemistry in Cambridge, UK. "JBS" was highly eccentric and awesomely gifted. As a student he mastered Greek, Latin, French, and German before fighting with manic bravery in the First World War. He had a habit of doing experiments on himself (there's a great photo in the departmental archive of him locked in a spasm after having injected himself with tetanus to "see what would happen"), he was a political activist, popular science writer, and broadcaster (in 1923, for heaven's sake, he pointed out that we would run out of coal as a source of power and should build a network of hydrogen-generating windmills!). He also made contributions to the study of enzymes and the field of genetics that are influential to this day. Regrettably, JBS died of colorectal cancer at the age of seventy-two.

Once he knew what was wrong with him, he composed a poem, not merely to make the point that you can't keep a good man down but also as an exhortation to others to consult a doctor as soon as you think you may have a problem. The first and last verses of his ode follow to set the tone. The entire epic is reproduced at the end of the book.

I wish I had the voice of Homer
To sing of rectal carcinoma,
Which kills a lot more chaps, in fact,
Than were bumped off when Troy
was sacked.
Yet, thanks to modern surgeons' skills,
It can be killed before it kills
Upon a scientific basis
In nineteen out of twenty cases.

My final word, before I'm done,
Is "Cancer can be rather fun."
I know that cancer often kills,
But so do cars and sleeping pills;

And it can hurt one till one sweats,
So can bad teeth and unpaid debts.
A spot of laughter, I am sure,
Often accelerates one's cure;
So let us patients do our bit
To help the surgeons make us fit.

In addition to the full version of Haldane's poem, at the end of the book you can also find the ode he might have written had he been alive today. There is also a glossary, an explanation of gene names, a list of sources of information, and links for website animations.

Robin Hesketh
Cambridge
t.r.hesketh@bioc.cam.ac.uk

PART I

HISTORY AND CAUSES

1

A SHORT STROLL THROUGH THE HISTORY OF CANCER

EARLIEST RECORDS

A FEW MONTHS BEFORE STARTING TO WRITE THIS BOOK I WAS PUFFing up a steep climb in the Drakensberg Mountains in search of one of the sites of rock paintings that are found throughout the region of what is now Kwa-Zulu-Natal. These artworks are the legacy of the San people who first moved to that part of the world about 8,000 years ago, and some of the paintings are over 2,000 years old. When we finally reached our goal, my first thought, apart from "thank heavens we can stop climbing," was what amazingly clear pictures these are of the animals with whom the San shared southern Africa. The most striking message, of course, was of the critical importance of hunting in the lives of those people. After a few minutes, however, I found myself musing on less obvious points. First that it was a bit surprising—given that securing your next meal was such a struggle, to say nothing of ensuring that you didn't become something else's dinner—that at least some folk had time to absorb the world around them and, moreover, come up with ways of recording what they saw. But then it occurred to me that, what with all the spearing and butchering that life entailed, these ancients must have built up quite a decent knowledge of anatomy, at least of wildebeest, gnu, and suchlike. Perhaps some far distant forerunner of Charles Darwin had sat on the very rock I was occupying and ruminated on how similar the layout of the bits was when you carved open the bodies of animals that looked completely different on the outside.

We shall never know, of course, because terrific artists though they may have been, the San had no written language by which they could pass on their knowledge. The first known written language is Sumerian, and records in cuneiform script have been dated as far back as 3000 BC. Cuneiform evolved through the practice of impressing shapes into clay, and pottery fragments bearing impressed markings have very recently been dug up in Pakistan that are estimated to be as old as 5,500 years. However, for the most ancient texts that tell a medical story, we have to turn to the papyrus records of the Egyptians. This method of recording may be almost as old as clay tablets, but its importance in the story of cancer emerged from the travels of an American called Edwin Smith, who visited Luxor in 1862. There he bought the manuscript that bears his name and that eventually came into the possession of the New York Academy of Medicine. Estimated to date from approximately 1600 BC, it was first translated in 1930 by James Breasted of the University of Chicago. The Edwin Smith papyrus is notable in at least two ways. First, particularly in its transcribed form as a pictograph, in appearance it is infinitely more beautiful than any twenty-first-century scientific paper. Second, it is in effect a set of very concise clinical reports, each presented under the headings Examination, Diagnosis, and Treatment, dealing with forty-eight cases. Almost all refer to patients who have suffered serious injuries ("a wound in his head penetrating to the bone of his skull"), indicative of evolution from the days of the San people at least in the capacity of humans to slaughter each other. However, two of the records refer to what Breasted translated as "tumors": "Instructions concerning tumors with prominent head in his breast: thou findest that the swellings have spread with pus over his breast, (and) have produced redness, while it is very hot therein, when thy hand touches him. An ailment which I will treat with the fire-drill" and "Instructions concerning bulging tumors on his breast (meaning swellings): findest them very cool, there being no fever at all therein. There is no treatment."

It seems probable that the first of these may have been an abscess—a build-up of dead cells at a site of infection—rather than a tumor, and that the symbol for a swelling was used for both. That would be consistent with the abscess being treatable with a fire-drill, assumed to mean cauterization using a red-hot iron, a substance with which the Egyptians were certainly acquainted, and a technique subsequently much practiced in Western movies. More notable from our point of view is the fact that the writer was sufficiently astute and honest to record that for "bulging tumors" there was nothing he could do, a state of affairs that arises to this day with some cancers despite the phenomenal progress that we will chart in our story.

The Egyptians also developed a considerable range of medicines; in the Edwin Smith papyrus, preparations of ostrich eggs and of honey are frequently recommended for the treatment of wounds. More perturbingly, their medicines often included minerals such as arsenic that we might expect to encounter in an Agatha Christie novel but not on the shelves of our local pharmacy. We now know that, quite apart from being a poison, arsenic can actually help cancers to develop. Nevertheless, both the Egyptians and the ancient Greek and Chinese physicians who used it to treat many conditions might have been on to something, because we will meet arsenic again when we consider recent developments in drug therapy for cancers.

THE GREEKS HAD A WORD FOR IT

Perhaps the most famous name in medical history is that of the Greek physician Hippocrates, often referred to as the "Father of Medicine," who lived around 400 BC. In fact, much of the story of his life that has come down to us is probably mythical, but he is credited with being the first physician to take a scientific view of disease as a natural process occurring within the body, as opposed to being visited upon us by some sort of magical, quasi-religious force. Certainly the writings attributed to him are a model of clarity in summarizing the most detailed observations following, albeit unknowingly, the pattern of the Edwin Smith papyrus. The field of cancer is indebted to Hippocrates for its name. He noted that tumors often had a high density of blood vessels, thought to resemble the limbs of a crab, the Greek for which gave us the word "carcinoma," the most common type of cancer in humans. We might also note in passing that Hippocrates was the first to use the Greek word "apoptosis," meaning the "dropping off" of petals or leaves from plants or trees, in a medical context. Hippocrates was actually describing gangrene, in which tissues become black and decay as a result of infection or restricted blood supply; we shall return to the subject of dying cells a little later in the story.

Galen, a countryman of Hippocrates, followed his footsteps some six hundred years later to become a prolific writer and radical surgeon—he was the first to treat cataracts and to employ taking the pulse as a means of diagnosis, and is generally credited with being the first to use the word "cancer," the Latin for "crab."

Avicenna (also known as Ibn Seena), the Persian physician, is thought of as a philosopher by virtue of his mastery of seemingly all science known at that time (around AD 1000), together with Islamic theology. He is credited with a large amount of written work, but his contribution to cancer was through a fourteen-volume *Canon of Medicine.* This became essentially the encyclopedia of medicine for the next seven hundred years not only in the Islamic world but also in Europe as a result of its being translated into Latin. It includes the first description of surgery for cancer and also a treatment using an extract from the plant *hindiba.*

LAYING THE FOUNDATIONS

A substantial period separated the observations of Hippocrates and his school from the first specific findings relating to cancer, which did not come until the eighteenth century and the dawn of the age of modern medicine. To a considerable extent, this long period reflects the fact that so often in science progress is limited by the tools available at the time; it was not until the seventeenth century that a significant surge of innovative intellectual activity signaled the onset of real progress. Perhaps the clearest indicator of the dawn of a new age was the founding of the Royal Society of London in 1660 under a charter granted by King Charles II. It evolved as a forum for a group of luminaries who had been exchanging information and ideas about science in an informal way and had become known as the Invisible College. The Royal Society is still going strong 350 years later in its elegant home just off The Mall in London. It is the British precursor of Academies of Science that many other countries have since set up and it is the oldest such learned society in the world.

Great strides in the history of humankind often reflect unique confluences of gifted individuals, and rarely can more brilliant galaxies have assembled than at the first gatherings of the Society. The founders included Christopher Wren, one of the greatest of all architects, and Robert Boyle, the founder of modern chemistry, immortalized by the fundamental law that bears his name (the one that says a gas fills less space the more you squeeze it—or, before any physicists have a fit, pressure multiplied by volume is constant). The Royal Society was shortly to publish Isaac Newton's *Principia Mathematica,* and Newton himself became president in 1703. One of the other founders was the outstanding Robert Hooke, a polymath if ever there was one. Hooke was born, the youngest of four children, into an ecclesiastical family living on the Isle of Wight. Fortunately, he was fascinated from his earliest years by all things mechanical, building a working wooden clock when still a youngster, and was thus saved from following the family tradition. This was a seriously good thing because, even among the bright sparks of the Royal Society, Hooke turned out to be rather special. He collaborated with the chemist Robert Boyle, who devised his celebrated law using vacuum pumps made by Hooke. He was a skilled surveyor and architect, coming up with a street plan for rebuilding London after the Great Fire (they didn't use it, which is why to this day London's chaotic layout reflects the Middle Ages rather than Hooke's grid-like vision). He was an astronomer, the first person to explain that light behaves like a wave and that as things get hotter, they expand. Moving into biology, he had a go at skin grafts and carried out the first recorded blood transfusion—on a dog. This prompted similar experiments on humans, mainly in France. However, as the subjects tended to die, in contrast to Hooke's dog, transfusions were abandoned until biochemistry was able to reveal the problem of matching blood groups.

For none of this, however, did Hooke become famous, and it required two further efforts to ensure his immortality. The first was working out his eponymous law of elasticity (how far a spring stretches depends on how hard you pull it). While Hooke's Law will forever be a fundamental of physics, his massive impact on biology followed from starting to play with microscopes. Simple lenses for magnifying objects had begun to be used by the late 1500s, and by 1624 Galileo Galilei had devised a compound microscope (one with two lenses), which is essentially the design used to this day. However, while Galileo's inclination was to point his magnifier at the stars, Hooke looked in the other direction, so to speak, and produced the first images of things in the natural world that are too small to be seen by the unaided eye. As it would be a further two hundred years before someone came up with the idea of photography, Hooke had to draw what he saw—and what a stunning artist he showed himself to be! His incredibly detailed pictures include a flea, a louse, and a gnat and he put them together in 1665 as the first scientific best seller, *Micrographia*, published by the Royal Society. Though not apparent at the time, its impact on the subject of cancer was immense. This came about in a slightly curious way because, in addition to various insects, Hooke pointed his microscope at plants, specifically at a piece of cork, for which the main source is the cork oak. Hooke's drawing shows vividly the regular structure revealed through his microscope that to him resembled the shape of a monk's cell. Thus for the first time was the basic unit of life revealed, and Hooke simply used the word "cell" to describe it (Plate 1).

While it was indeed many years before it became clear that cancer was a reflection of something going wrong with cells in animals, Hooke's observations and the other great scientific events of that time opened the eyes of all with an interest in the natural world to the notion that, by observation and intelligent deduction, it might be possible to make sense of what hitherto had been explained only by religious mysticism.

This realization prompted a gradual trend to thoughtful analysis of cause and effect, one of the first exemplars being an Italian, Bernardino Ramazzini. He noted in 1713 that nuns rarely fell victim to cervical cancer and yet were particularly prone to breast cancer and concluded that this might have something to do with their lifestyle. Of course, his observations reflect the fact that cervical cancer most commonly results from the transmission of a virus during sexual contact and that the hormonal changes associated with pregnancy are a strong protective factor against breast cancer. As nuns are generally held to avoid both these activities, they could perhaps be viewed as losing on the swings what they've gained on the roundabouts.

Shortly thereafter, in 1761, Dr. John Hill of London sparked a long-running debate by concluding that snuff inhalation, popular in those days among the male gentry, might cause nasal cancer. The snuff saga persisted for over two hundred years until, in 1985, the Cancer Research Campaign brought its authority to bear by stating "there is no evidence of any association with cancer or other health risk in the snuff produced in this country. For this reason, snuff seems an entirely acceptable substitute for cigarette smoking and could be recommended for addicted cigarette smokers since, if they could substitute snuff taking for cigarette smoking, they would greatly reduce the risk to their health." Well, that piece of advice could scarcely be said to have made a major impact, but it did make the point that, in a sense, the good doctor Hill was on the right lines in thinking that something you inhaled might give you cancer. However, the critical difference between snuff inhalation and smoking is combustion—it's burning tobacco that releases chemicals that cause cancer, over forty of which have now been identified in cigarette smoke.

A more notable physician of the time was Percivall Pott, mainly because he tended to be right. Pott worked at Saint Bartholomew's Hospital in London, and to this day his name flits before the consciousness of medical students because it is given to a particular type of ankle fracture and also to a form of tuberculosis of the backbone. Seemingly, Pott broke his leg when he was thrown from his horse while making a house call. He broke both the bones in his lower leg and they penetrated the skin. In those days the recommended treatment was amputation, a strategy on which Pott wasn't too keen, and instead he tried binding up the limb and resting it. He made a full recovery, spending his convalescence writing a treatise on the management of fractures and dislocations. This story indicates that Pott, as well as being a gifted surgeon, was an able scientific observer, as he showed by spotting that chimney sweeps often succumbed to cancer of the scrotum. In 1775, rather than using mechanical aids, chimneys were swept by boys shinning up the vents, a job that they often carried out naked. Those who were astute or coy enough to wear a leather garment about their nether regions were protected and Pott, comparing the incidence of cancer with the modesty of the sweeps, concluded that the cancer was caused by soot accumulating in the folds of the scrotum. This was the first identification of an occupational exposure to cancer-causing agents, and it led to other such risks being recognized. However, soot, like tobacco smoke, is a complex mixture, and it took until 1930 for Ernest Kennaway to isolate the first specific chemical that can promote cancer (i.e., it's a "carcinogen"). The stuff Kennaway pulled out was a "fused hydrocarbon," which is the main carcinogen in both soot and tobacco, though we can also get doses of them from forest fires and asphalt. We now know that quite a number of these "promoters" are converted into harmful forms only once they get inside our bodies: the more deeply you inhale, the more this happens—and the more likely you are to get lung cancer.

Nowadays the immediate reaction of a scientist to Pott's observation would be "let's do a controlled experiment in which we put soot on skin and see if it causes cancer." History doesn't record whether Pott floated the idea of an alternative career as a laboratory animal to chimney sweeps during the reign of George III, but we do know that the first controlled experiments to investigate cancer were not recorded until the twentieth century. Notable among these was the first demonstration that substances derived from coal can induce cancer. This was in 1915 by Katsusaburo Yamagiwa and Koichi Ichikawa at Tokyo University, who applied coal tar to the ears of rabbits every two or three days for over three months, whereupon skin cancers did indeed form. As we shall see later, however, it is not just chemicals that can cause cancer: shortly before this experiment, Jean Clunet had shown that another way of causing skin cancer, this time in rats, is by exposure to X-ray radiation.

A FIRST LOOK AT CELLS AND MOLECULES

These inquisitive souls laid the foundations for studying the causes of cancer, but, in addition to their efforts, no less astute observers were setting out on the road that would lead to the analysis of cancers at the level of the molecules involved. Among the first of these was René-Théophile-Hyacinthe Laënnec, who invented the stethoscope in 1816, seemingly as a way round the problem of listening to the heartbeat of a young lady whom he described as having a "great degree of fatness." Although she doesn't sound too appealing, the lady may have been lucky because obesity is associated with a lower density of breast tissue (i.e., less fatty breasts) and that carries a lower cancer risk—a point we'll return to. For the moment, however, Laënnec's cancer connection was as the first person to describe melanoma, the abnormal growth of skin pigment cells, and to note that such growths could spread to the lungs. Some nineteen years earlier, the Scottish surgeon John Hunter had carried out the first recorded treatment of melanoma. He removed a secondary growth that had formed after cells had traveled to the new site from a primary tumor, although it was 150 years before the excised tumor was actually shown to be a melanoma.

In 1820 a Stourbridge doctor, William Norris, extended the observations of Laënnec and subsequently described a family in which individuals from several generations had developed the same form of cancer. This inference that some families might be predisposed to cancer was extended by the extraordinary French physician Paul Broca. Broca was not only a leading surgeon but an outstanding anatomist, and his delineation of regions of the brain, particularly those regulating speech and smell, ensured that his name will forever be even more familiar to students of medicine than that of Dr. Pott. In 1866 Broca suggested it might be possible to inherit breast cancer. He'd looked at his wife's family tree and noted that ten out of twenty-four women, spread over four generations, had died from that disease and that there had been cases of other types of cancer in the family as well. This large proportion was not, he believed, mere chance. We know now, of course, that a changed (mutated) form of a gene (a unit of heredity), passed from generation to generation, was almost certainly responsible for the suffering of this family.

This is our first encounter with the idea that inherited changes can affect the way cells behave. We'll come to what all this means in terms of molecules a bit later, but for the moment, a perhaps slightly surprising analogy might be useful. We noted that the founding of the Royal Society in 1660 was a seminal event in the story of human civilization, but it will doubtless have crossed your mind that something even more important occurred in the fifteenth century when an inventive German by the name of Johannes Gutenberg came up with movable-type printing. From the Egyptian scrolls through Hippocrates until that point, all the knowledge, stories, and legends recorded by mankind were the result of the Herculean efforts of scribes who, with endless patience, engraved, copied, and re-copied everything that had come down to us. Inevitably, each copy differed: tired eyes made mistakes, tea breaks (even monks must have had them) meant lines or even paragraphs were skipped, smart minds put in "improvements," and so on. To illustrate, here are two examples of the same line from different versions of John Lydgate's *Kings of England*, a poem about William the Conqueror and his successors:

Buried at cane this saith the croneclere
And is buried at Cane as the Cronycle sayes

The evolution of manuscripts turns out to parallel the copying of DNA. The work of replicating DNA so that copies can be given to new cells is carried out by proteins. As we shall see shortly, proteins are wondrous things but nevertheless, they have frailties. They too make mistakes and sometimes they become perverted (mutated) and then make even more errors. This parallel is more than notional—the same math models that track errors in manuscripts can be used to follow changes in DNA and draw those family trees that show how closely related humans are to chimps or bananas. The useful image to keep in mind is of each letter in the poem being one of the four in the DNA code. The two lines clearly *mean* the same thing, but in a DNA code the many differences are "mutations." These may have no effect but, as we'll see shortly, even one change may mean life or death.

Of course, at this point the concept of genes had not been invented, so to speak, but as the nineteenth century drew to a close, a number of people were beginning to look closely at what we would now call the genetic material of cells—the individual chromosomes that together make up the store of DNA. Individual chromosomes were first visualized during the process of cell division (when a parent cell divides to form two daughters) in the middle of the nineteenth century, and they turn out to be surprisingly easy to see. All you need is a simple light microscope because, as cells divide, their chromosomes become highly compressed in preparation for being split between the two new cells. When these "condensed" chromosomes are spread on a glass slide, their gross features are clear, even at the comparatively low magnification of 100 times.

While Broca was making his perceptive observations and deductions, just across the border in Switzerland a character with whom he had a good deal in common was also discovering something amazing, although his efforts remain to this day largely ignored by the scientific community. Carl Christoph Vogt was born in Giessen in Germany and was taught for a while by Justus von Liebig, one of the most influential chemists of all time, which must have been a life-changing experience for him. Liebig had become a professor at the age of twenty-one, having, no doubt, alarmed his parents by announcing that he wanted to be a chemist before such things had been invented. Having been packed off to be an apprentice apothecary, Liebig set up his own chemistry lab in the shop and, so the story goes, terminated his apprenticeship and almost himself in an explosion of fulminates—chemicals that are used as the explosive in detonators. In fact, this propelled him into a university career that brought forth organic chemistry as a central part of physiology and pathology, showed that nitrogen is an essential nutrient for plants, and established for the first time a rigorous system for teaching chemistry. By coming up with a method for making beef extract, he also gave the world Oxo bouillon cubes. After his time with Liebig, Vogt spent most of his scientific life in Switzerland, and it was in Neuchâtel while studying tadpoles (specifically those of the midwife toad *Alytes obstetricans*) that he made what must have been a very puzzling observation: as the tadpoles developed, some cells seemed to dissolve and vanish. "The nucleus gradually disappears and you are left with rather homogeneous secondary cytoplasm in which you find dispersed cellular vacuoles [hollow cells] and sometimes you can see the rest of the nuclei as shadowy flecks amongst them," wrote Vogt. That was a serious bit of experimental observation, given the means available in 1842 and the difficulty of following the disappearance of something. In fact, this problem wasn't fully overcome until time-lapse cinematography could be used to record the activity of cells over extended periods.

The most significant figures in the early stages of cancer molecular biology were two German scientists, Hansemann and Boveri. In 1890 David Paul von Hansemann reported his observations that chromosomes within human tumor cells could divide abnormally and came up with the term "anaplasia" (meaning "to form backward") to describe the loss of normal cell characteristics that has occurred in a tumor cell—and, as we'll see later, this really lies at the heart of cancer, when cells cease responding normally to the world around them and reproduce themselves without restraint. Theodor Boveri, who was born in Bamberg and studied the humanities before switching to biology, had already showed that heredity is carried by chromosomes, though most of the leading geneticists of the time, including the American Thomas Hunt Morgan, didn't think too much of this idea. However, in the way that science often has of humbling people, it was Morgan himself, doing experiments on flies very similar to the earlier experiments of Gregor Mendel on peas, who confirmed in 1910 that genes do indeed reside on specific chromosomes. Mendel was the celebrated monk-turned-scientist, born in what is now the Czech Republic when it was part of the Austrian Empire. By tracking things like flower colors in the monastery garden at Brno, he was the first to show that inherited traits come in little packets—genes. Funnily enough, despite his initial skepticism, Morgan is perhaps better remembered than Boveri because J. B. S. Haldane, whom we met in the Preface, suggested that his name should be used as the measure of distance along chromosomes (in practice, it's centimorgans).

Meanwhile, as Morgan was gearing himself up for a U-turn, Boveri, building upon the work of Hansemann, pointed out in 1903 that a fault in the machinery controlling cell division might produce cells with an abnormal number of chromosomes or damage them in some way—in turn leading to cancer. The essential idea was that a normal cell turns into a tumor cell when something goes wrong with its chromosomes. Boveri noted that there could be "countless" different abnormal chromosome combinations but that almost all of these would be fatal for the individual cell. In other words, only very rarely would a cell acquire a set of genetic changes that would allow it both to survive and to develop the properties that make a tumor. Since Boveri's time microscopy has become cleverer, and you can now tag each chromosome pair with its own color. Using the same tags to look at chromosomes from cancer cells provides stunningly beautiful confirmation of Boveri's insight. Imagine each normal chromosome as a sausage, each with its own color: slice and shuffle, then glue the bits together to reconstruct chromosomes. The result is spectacular banding patterns in the tumor chromosomes caused by the shuffling of large chunks of DNA that goes on as almost all cancers develop.

Boveri, who was a zoologist, made his deductions from working almost entirely on the eggs of spiny round creatures called sea urchins that live on the sea bed (an attractive system because it's fairly easy to get large numbers of fertilized eggs, they're rather transparent, and all the cells divide at the same time as the embryo develops). He also used a type of worm and, as we shall see, he was not the last person to make decisive contributions to human cancer biology through seemingly humble organisms.

Boveri's name will always be associated with that of the Kansas-born Walter Stanborough Sutton because together they established the basic picture of what happens when a cell is fertilized and starts to divide, which teenagers learn nowadays in school biology classes. Every normal cell has *two* sets of chromosomes, and before it divides, the chromosomes must be duplicated so that each daughter can have a complete set. However, egg and sperm cells are formed so that each has only *one* set of chromosomes: therefore, when they fuse during fertilization, you get one cell with two complete sets of chromosomes. All of which provided the first clues that chromosomes were the carriers of inheritance. Boveri also suggested that normal cells might have a built-in system that keeps them from dividing until they receive some overriding signal, a remarkably prescient thought some sixty years before we started to work out that a cell does indeed have to pass a series of "checkpoints" before it can divide.

CUT, BURN, AND HOPE

We have seen that, at least by the time of the ancient Egyptians, attempts were being made to treat cancers. As science moved into the nineteenth century, progressively more daring and innovative methods were pioneered on the direct intervention front and, very gradually, these began to go hand-in-hand with the first unsteady steps toward the study of cancer at the molecular level. We have already encountered John Hunter performing surgery on a melanoma, but there are two or three other figures whose efforts should feature in a summary of major landmarks. The first is Christian Albert Theodor Billroth, a German who helped establish the foundations of modern surgery. He carried out the first removal of the larynx for cancer in 1874, the first successful surgical treatment for stomach cancer, and the first removal of a bowel cancer. Billroth had earlier realized that there was a relationship between benign growths (polyps) on the lining of the bowel and cancers, a point to which we will return later. In addition to his surgical skills, Billroth was a gifted musician who actually studied music for a brief period at the University of Greifswald before becoming engrossed in medicine. In 1859 he became professor of clinical surgery at Zurich, where he became lifelong friends with Johannes Brahms, who rendered Billroth unique among cancer scientists by dedicating a string quartet to him. It was perhaps fortunate that Billroth predeceased Brahms by three years, thus being spared the painful irony of seeing his friend die of liver cancer.

The second major surgical figure is George Thomas Beatson, one of the leading clinicians of his day, who shortly after he graduated became house surgeon to Joseph Lord Lister. He is principally known for his studies on female cancers, and he carried out the first operations in which the ovaries were removed as a way of treating women with advanced breast cancer. This approach evolved from his realization that one organ could exert control over another by means of chemical messengers released into the circulation. The messenger in question was the primary female sex hormone estrogen, although it was to be thirty years before it was isolated. In a slightly different approach, Beatson treated breast cancers with extracts from the thyroid. Although he could not identify the agents he was dealing with, he was convinced that unraveling the chemistry, that is the molecular processes involved, was essential for an understanding of cancer; his efforts laid the basis for today's hormone therapy. It is wholly appropriate therefore that he is best remembered through the names of two of the world's leading research institutes, the Beatson Institute for Cancer Research and the Beatson Oncology Centre in Glasgow.

Fifty years later, the work of Beatson was complemented by that of the Canadian Charles Brenton Huggins. His specialty was the urinary tract, and he showed that removal of the testes could stop the growth of prostate tumors and indeed make them disappear. Testosterone (the main male sex hormone, first isolated in 1935) reversed this effect and in dogs prostate tumor growth could be switched on or off by administering or withholding the hormone. We've just met the corresponding female sex hormones, estrogens, which are important in men too (they play a role in sperm development), even though women usually have more of them. By the 1940s estrogen had been purified (from the urine of pregnant women) and shown, when injected into male mice, to cause breast tumors. These advances enabled Huggins to show that estrogens had the opposite effect of testosterone on prostate tumors. That is, estrogens could block the growth-promoting effect of testosterone on prostate tumors and provide an effective treatment without the requirement of castration.

The work of Charles Huggins marked a new era because, until the middle of the twentieth century, cancer treatment had essentially developed empirically—by trial and error, to put it more crudely—uninfluenced in any direct way by systematic scientific studies carried out in laboratories. There had, of course, been huge contributions from great scientists, most notably Louis Pasteur and Karl Landsteiner. Pasteur, together with Robert Koch (who isolated the bacteria responsible for tuberculosis, anthrax, and cholera), showed that microorganisms caused many diseases, discoveries that eventually led to the introduction of antiseptic surgical techniques. Landsteiner established a system for the classification of blood groups in the early 1900s and thus provided a rational basis for blood transfusion based on matching blood types. Pasteur, Landsteiner, and Koch's great discoveries were, however, somewhat indirect in their impact on the actual practice of medicine. Huggins produced the first example of fundamental scientific research being directly transferred to medical practice. He acknowledged that his triumph was possible only because of the laboratory labors of many others with the words, "These wonderful discoveries provided the Zeitgeist for our work." For his efforts in launching the field of chemotherapy Huggins shared the 1966 Nobel Prize in Physiology or Medicine; we shall meet his co-recipient later in the story.

William Sampson Handley, once described as "the most fertile brain in British surgery," was particularly noted for his work on the spread of breast cancer when he worked at the Middlesex Hospital. He developed methods for killing residual cancer cells after breast surgery by chemical treatment and for defining limits for the excision of skin melanomas to prevent recurrence of the tumor at the primary site. The notion that chemicals could kill cancerous tissue goes back at least to the Egyptians and their use of arsenic. In the nineteenth century Sir Humphry Davy, while showing that acids reacted with metals to form salts, noted that zinc chloride had a strong caustic action on tissues. The first description of the clinical use of zinc chloride was in 1910, and it is used nowadays in a variety of commercially available pastes and salves for treating superficial skin cancers. Fused silver nitrate, which can be compressed into sticks that look a bit like gray eyeliner, can also be used as a cauterizing agent and is a quick and effective treatment for mouth ulcers. However, it should be noted that neither of these chemicals is at all specific for cancer cells: they kill normal tissue with equal efficiency.

Zinc chloride has played an additional role in the history of cancer surgery because of a chance observation by the American physician Frederic E. Mohs. He noticed that when injected into tissue, it is a very good preservative. What's more, if you then cut the tissue into thin sections, the zinc chloride highlights the detailed structure of the constituent cells so that you can distinguish normal from tumor cells. This was the precursor to the method of fixing, sectioning, and staining tissues that is now used worldwide. However, Mohs turned his finding to a more specific use by recalling Handley's concept that it is critical to define the limits of a tumor if surgery is to be effective and not leave the patient with a high risk of recurrence owing to the growth of unremoved tumor cells. Mohs's strategy was to inject zinc chloride into tumors before cutting them out in steps, examining each fragment as he went, until he was satisfied that no more tumor cells were being removed. The method was subsequently refined so that the tumor sample was removed from the patient before any chemical treatment. The technique of Mohs's micrographic surgery thus involves microscopic analysis of tumor samples that have been removed from the patient and stained to permit detection of cancer cells. The process of surgical removal, slicing, staining, and microscopic analysis is repeated until no further cancer cells are found. For the most common type of skin cancer, this method gives a 97 to 99 percent five-year cure rate.

Our final surgical hero is William Stewart Halsted, an outstanding American surgeon who is important in the present context because he performed the first radical mastectomy for breast cancer. Halsted developed a number of other surgical procedures and is also credited with coming up with the idea of surgical gloves, an advance prompted by the requirement to protect the hands of his scrub nurse (who responded to this concern by marrying him). Halsted's treatment was radical because it involved removal of the breast and underlying muscle and also the lymph nodes of the armpit to which breast tumors often spread. In fact, in the early years of the twentieth century, breast cancer surgery could be even more radical and extend to hip or shoulder amputation. That seems to modern eyes to verge on the barbaric; nowadays any kind of extensive surgery is required only rarely, but for a long time Halsted's pioneering method was the only one available, and his concepts influenced the treatment of breast cancer in particular for much of the twentieth century.

Such desperate measures were, of course, a reflection of the knowledge of the time. Halsted and his contemporaries knew that the killer in cancer was almost always secondary tumors formed from cells that had seeded from the original primary growth. Indeed, he was very aware of what we're so often told today: the earlier a cancer is treated, the better. What's more, he noted that secondaries could show up years after a primary tumor had been removed, there having been no sign of any secondary disease at the time of surgery. But the big problem that the surgeons faced was having no idea of how tumor spread occurred. Halsted felt that there was continuity between the primary and secondary and, because breast tumors often spread to the bone, that was a good reason for trying ever more extreme surgery as a means of preventing recurrence. All of which was not unreasonable and, as the glimmerings of how tumors spread through the body are only just emerging as we enter the twenty-first century, we can but have sympathy for those who grappled with what to do over one hundred years ago.

In his writing Halsted comes across not merely as an innovative surgeon but also as a rigorous scientist and a concerned doctor. In discussing the difficulty of "curing" breast cancer once it has spread, he demands "incontrovertible proof" but concludes that "even if the microscopic findings were confirmed by an able pathologist I should still feel that an error might have occurred, for example, in the labeling of the specimen." No detached, God-like figure was he, but one who recognized that even the best-run outfits can perpetrate what nowadays might delicately be called a "booboo." On the matter of early detection of tumors he raises a question that is just as pertinent today: "Shall we let women know that a dangerous process may be going on which they cannot detect, and keep them in a constant state of apprehension, or shall we encourage them to seek 'expert' advice which may be insufficiently expert, and expose them to the annoyance of repeated and useless examinations, each of which for only a brief period, if at all, would bring a measure of reassurance?"

These heroic individuals, through their technical genius and perception of how the human body works, established the basis of cancer both as a scientific discipline and as a field of clinical medicine. It remains true that the first method of treatment for many cancers is surgery, which is still the most effective method for enabling people to survive the disease. In a crude sense, therefore, "cut, burn, and hope" does sum up the foremost strategy. However, this aphorism undervalues the efforts of these men in laying the foundations of clinical methods that have become increasingly sophisticated. For example, surgeons now specialize in specific cancers using increasingly refined technical aids and this trend toward ever greater, focused expertise will continue as novel methods for defining the geometry of tumors and also for removing them are developed.

THE MOLECULES THAT MAKE US

Around the beginning of the twentieth century, a new word began to edge its way into the scientific lexicon—biochemistry—referring to the study of what might grandly be called the chemistry of life, namely of proteins, sugars, and fats and, of course, DNA—the molecules that make living systems. Biochemistry really began to emerge with the dissection of the major metabolic pathways that break down what we eat to provide energy and that build the molecules of which we are made. In the period following the First World War, the pathway that breaks down the sugar glucose to release its energy, called glycolysis, was defined. The complete breakdown of glucose so that all its energy is released requires a second pathway, worked out by Hans Krebs in 1937 and often called the "Krebs cycle." Krebs was the first of a prodigious stream of European scientific brilliance that passed through the Biochemistry Department in Cambridge from 1933 onward. This was largely owing to the efforts of its head, Frederick Gowland Hopkins, who was so aware of political developments that as early as 1933 he set up the Academic Assistance Council to find places for

European refugees. Krebs was one of the first victims of the ban on Jews teaching in German universities or practicing in university hospitals. By early April 1933, Hopkins had written to Krebs offering him shelter, and he left Freiburg in June for Cambridge.

An indication of how important these molecular approaches to living systems were to be for cancer was not long in coming. By the 1920s Otto Warburg had noticed that something odd happened to metabolism in cancer, and he showed that tumor cells get most of their energy from glucose using the glycolytic pathway, despite the fact that it is less efficient than the Krebs cycle.

Warburg was part of an amazing scientific galaxy in the period from 1901 to 1940 in which one out of every three Nobel Prize winners in medicine and the natural sciences was Austrian or German. Born in Freiburg, he completed a PhD in chemistry at Berlin and then qualified in medicine at the University of Heidelberg. Fighting with the Prussian Horse Guards in the First World War, he won an Iron Cross and followed that up with the 1931 Nobel Prize in Physiology or Medicine for showing that cellular respiration, that is, oxygen consumption, involves proteins that contain iron. However, he made so many contributions to biochemistry that he was nominated three times for the prize.

In the 1940s the Canadian Oswald Avery showed that genes and chromosomes are made of DNA and then, in 1953, came arguably the most famous discovery of all when Watson and Crick deduced that DNA forms a double helix (Plate 2). Francis Crick, having studied physics at University College London and worked on magnetic and acoustic mines during World War II, gravitated into biological research at Cambridge. In 1951 he was joined by the Chicago-born James Watson, who had just acquired a PhD from Indiana University. Most contemporary accounts recall Watson as not being short of self-confidence while Watson himself famously opened his account of the events that followed (*The Double Helix*) with the words "I have never seen Francis Crick in a modest mood." Over the next two years their collaboration identified the structure of DNA, the molecule of life, in what has been described as "the greatest achievement of science in the twentieth century." Perhaps unsurprisingly, the saga has become the most discussed and controversial episode in scientific history. Thirty years later Watson was to muse that he didn't "think the whole thing would have worked if the Cricks hadn't cooked so many meals for me, or made me feel at home in Cambridge by seeing that I didn't cut my hair for quite a while. And then seeing that I wore a tie. And that I got an English suit. And giving me the good advice that I shouldn't look like an American. I followed their rules to the point where it made it difficult sometimes to go home." Touchingly and somewhat at odds with the general perception of his character, he concluded "I couldn't have got anywhere without Francis. . . . It could have been Crick without Watson, but certainly not Watson without Crick." By 1955 Fred Sanger had obtained the first sequence of the individual units that make up a protein, and the secret of how proteins are made using the DNA code was being revealed. This eventually led, in 1977, to the invention of a method for sequencing DNA, again by Sanger. It was this technical feat that, in turn, led to the launch of the project to sequence human DNA in its entirety in 1988, in which James Watson played a major role, and to the triumph of its completion in 2003 (Plate 3).

That continuing story of the application of scientific method, predominantly biochemistry but increasingly engineering and mathematics, and how it has provided a rational basis for cancer—how we view it, how we treat it, and how it is driving the development of diagnosis and therapy—is really what this book is about. Inevitably, the following chapters that bring us up to date will rely almost entirely on discoveries made in the last twenty years. Being about science, there will be quite a lot of "facts," but the human endeavors behind the emergence of knowledge remain as fascinating as ever, and we will meet some of the scientific descendants of these great men as the story of the impact of science on medicine and surgery unfolds.

WHERE ARE WE GOING?

After this brief flirtation with history and before we launch ourselves at cancer, it might be appropriate to pause for a moment—and ask "why?" Why should we confront this disease and what might we gain from doing so? It's true that it will kill one in three of us but, look on the bright side, that means the other two of us will die from something else. Either way, had you asked these questions at any time up until 1953, the sensible response would have been, "try not to think about it because there's no point." Before then, although we knew that cancers were the most complex diseases to afflict mankind, that they could target pretty well any part of the body, and that neither age nor sex, race nor habitat, offered protection, we knew absolutely nothing about their causes at the level of the affected cells. So a sensible approach was perhaps for scientists to continue collecting information about the disease, especially at the molecular level, but for the general population not to worry too much about trying to make head or tail of it all. That somewhat defeatist view would in no way have minimized the efforts made over the centuries by gifted and skillful surgeons trying to alleviate our suffering. It would merely have been a realistic assessment of the situation. Mind you, when you think about some of the radical cancer "cures"—removal of entire limbs and the like—that were practiced into the first half of the twentieth century, you might have been inclined to opt for being left alone to suffer in relative peace.

What happened in 1953 was, of course, the description by Watson and Crick of the structure of DNA and, although it didn't exactly happen overnight, that started the revolution in biology that continues to this day. Very soon folk were identifying genes that did abnormal things in cancers. You could argue that almost any bit of the cellular machinery that makes us tick could have a hand in cancer development, and certainly it wasn't long before the number of known major cancer-associated genes ran to hundreds. The quite staggering technical advances that have been part of this revolution now enable us to look at the complete pattern of molecular changes associated with cancers, and it seems likely—astonishingly—that essentially every cancer is different in the fine detail of these changes, even though subgroups of cancers may be indistinguishable to clinicians or pathologists. At first pass, this seems even more daunting. We have been swept by what one can only call a tidal wave of facts because scarcely anything other than the tsunami metaphor will suffice. However, there may be some glint of sunshine above the surf. Firstly, the very fact that there are so many molecular changes associated with cancers in principle offers a large number of targets for therapy. Secondly, what Sir Peter Medawar called "the ballast of factual information" and "the tyranny of the particular" may not be quite the sandbagging it sometimes seems. Medawar's view was that you might have to go through this sort of phase in science but that gradually, from the welter of facts, would emerge some kind of pattern, consensus, or general principles by which nature works (Plate 4).

Medawar was a great scientist. He shared the 1960 Nobel Prize with Australian Frank Macfarlane Burnet for showing that when organs were transplanted from one animal to another, they were rejected as "foreign" by an immune response that was "learned" by the animal during early development and that could be overcome. All the subsequent organ transplantations and the methods for suppressing the immune response stemmed from their findings. However, Medawar was also one of the finest of scientific writers and most gifted of communicators about science, about how it worked, why it was such fun and so important, and about the broad philosophy of the subject. Any of his writings are informative, witty, and brilliant examples of the writer's art, from *The Strange Case of the Spotted Mice and Other Classic Essays on Science* to his review of James Watson's *The Double Helix*. Quite apart from all this, I owe Peter Medawar a huge personal debt because he gave me my first proper job in the UK after I returned from North America, at least if you count doing a PhD as an "improper" job. That was after he had become director of the National Institute for Medical Research in Mill Hill in 1962, thus achieving just about everything you could wish for in science. Nevertheless, he reckoned that running a major institute and all his external commitments could be fitted into about half the week and that Tuesdays and Thursdays were still for his own bench work. One of my most vivid impressions was the regular spectacle of Medawar leaving his office on the ground floor, turning right, and bounding up the stairs three at a time to get to his first floor lab. It's difficult to think of a better image to represent all that makes science such a wonderful pursuit.

So the question is, can we apply to cancer Medawar's general view that through a mountain of scientific data a light will eventually shine? I believe we can, and I will try to show that there are some basic patterns of how cells work about which we now have a reasonable understanding—and that we can note, but perhaps set aside, some of the complexity as being less a driving force and more a being carried along on the wave, so to speak. From this emerges a view of how we are beginning to design on a rational basis—that means we think we know what we're doing—methods for treating cancers. The incredible technical developments of the last few years mean we can now identify all the mutations in individual cancers. In principle, that in turn means we can confront cancers with combinations of treatments that are individually tailored to their stage of development. All of that has happened in less than fifty years. So it's taken rather a large part of human evolution to make any progress at all on the cancer front but, having started, the vista that is opening is truly stunning.

2

COUNTING CANCERS AND COMPARING CAUSES

IT HAS TO BE TRUE THAT IF YOU TOOK A GLOBAL VOTE ON WHETHER cancer was a "good thing," 99.9 percent would tick the "bad" box. There might be a few scientists in there too who might be voting for their job but, pressed on the altruism front, they could very reasonably claim that the pursuit of cancer has been a uniquely powerful driver in revealing how animals work. As we shall see, there is no aspect of the molecular clockwork of life that isn't involved in these diseases and, as answers have emerged to the puzzles of cancer, many of the secrets of life have been unveiled.

Given that human beings are all much the same biologically speaking, the fact that there is a wide range of types of cancer suggests that we have considerable control over many of the causes. After our brief brush with numbers, we'll see that causes can indeed be divided, very broadly, into two groups: those we can do something about and those we can't, and the sensible approach is to forget about the latter but give a bit of thought to the former. In following this advice, we see that there are some things we should do and some that we should not. One "to do" is "be lucky"—that is, live where there is a decent standard of hygiene so that you minimize the risk of infections that can cause cancer. Another is "be careful"—don't run unnecessary risks of catching viruses from others and, if you're fair of skin, don't lounge around in the sun without protection. "You are what you eat," we're often told. Of the thousands of chemicals that enter our mouths every day, some are essential for our survival. Some of the others should come with a warning—but which? You've probably spotted that scientists often perch firmly on the fence when it comes to questions to which we'd really like to have a "yes" or a "no." Why do they do that? Why is working out what is bad for us such a problem? And why, when it comes to treatment of major cancers, do we see such variation in outcome?

A QUICK LOOK AT NUMBERS

Most people know that one in three of us will get some form of cancer, so you don't need to be a math wizard to see that plunging into cancer stats is a bit like giving someone a large sandbag and inviting them to whack you over the head. Even so, it's worth a bit of pain because, as we shall see, this turns out to be an unusual example of numbers being both interesting and instructive. Cancer is so powerful because the numbers really are overwhelming. Every year more than 12 million people world-wide discover they have a cancer and each year over 7 million die from one. The major killers are lung cancer followed by stomach, liver, bowel, and breast. The figure of 1.4 million lung cancer deaths accounts for just over 18 percent of all cancer deaths: for breast cancer it's about half a million (6 percent). These figures are for 2008, the most recent year for which global figures have been compiled. If you think that's a bit slow when it comes to spreading bad news, a glance at the amazing website of the Union for International Cancer Control will convey some idea of the massive labor involved. As you might expect, the pattern for new cases is broadly similar to that for mortality—with a little bit of shuffling—so that the top six are lung, breast, bowel, stomach, prostate, and liver. Specific causes and available treatments account for these differences. Most liver cancers arise from infection by viruses—something that rarely happens in the developed world but is relatively common in, for example, Africa. The disease is essentially untreatable, hence its sixth position in the incidence league but it is third as a killer.

Americans and Britons will be unsurprised by these statistics because almost everyone knows that lung, breast, and bowel cancers are in our top five

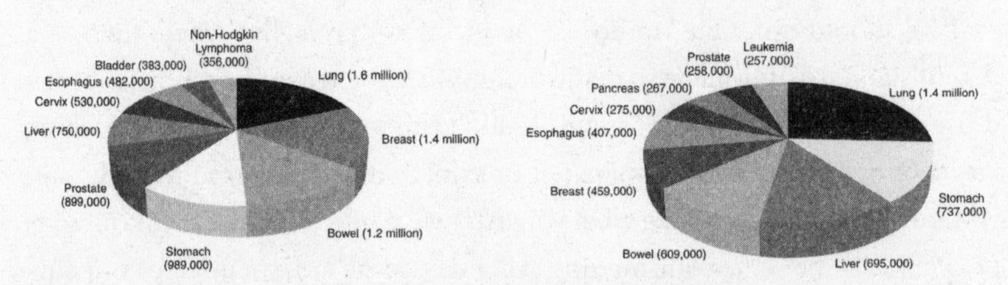

Top ten: world cancer new cases (left) and deaths (right) in 2008

of both new cases and cancer deaths. The other two Big Five killers in the US, the UK and most of the developed world are prostate and pancreatic cancers, eighth and ninth in the global rankings. The main message is that, wherever you live, the dominant forms are lung, breast, and bowel cancers: the overall pattern is fairly similar across the developed world but shows significant differences from the global picture. Stomach and liver cancers are in the world top six but don't get into the US and UK top ten. There are other big variations but for the moment the point is that, bearing in mind that cancers are caused by mutations and human beings are about 99.9 percent identical as far as their DNA goes, these global differences tell us two useful things. The first is that our lifestyle (where we live, what we eat, etc.) plays a big role in cancer. The second is that the best treatments currently available are quite good for some types of cancer and almost useless for others.

For one "lifestyle" example, consider the absence of stomach cancer from the US/UK lists. 'Twas not ever thus: go back eighty years in the US and it was the biggest cancer killer, even though the total death toll in 1930 was only about 120,000. Nowadays nearly two-thirds of stomach cancers occur in the developing world, but there too the incidence is declining, as it has done in the US. It's not really clear why this form of cancer is in decline, but the increasing availability of fresh fruit and vegetables and of meat preserved by refrigeration rather than by salting may be helping. Nevertheless, the estimate is that 10,570 will die of stomach cancer in the US in 2010.

Generally speaking, the global cancer variation, though significant, is not enormous, but there are some major differences. Countries with the lowest rate of new cases (e.g., Gambia) have about five times fewer than the US for all forms of cancer. Some specific cancers show still greater variation: there are about 300 new skin cancer cases in parts of Australia for every one in Kuwait. The mortality rate (deaths per 100,000 in the population) in India is about half that of the UK—mainly reflecting the average life expectancy (64 in India and 80 years in the UK). The variation within Europe is less dramatic. The UK is around the middle of the table with Hungary being top (the bad end) and Finland doing better than everyone else; the Finnish death rate is about 78 percent that of the UK.

On the treatment front, note several common US/UK cancers that aren't in the top ten mortality stats—cervical cancer and melanoma, for example—indicating that these can, by and large, be successfully managed. For others (lung, pancreas) the charts show that, for all the effort, cash, and ingenuity that has been directed at them, some cancers remain essentially intractable. Overall, the balance of the last twenty years in the developed world has seen a steady increase in the five-year survival rate, albeit with some countries (US and Finland) doing rather better than the rest. Japan is also notable because its death rate from cancer has been consistently some 20 percent lower than that of other developed countries. Because it comes up so often, we should note that the "five-year survival rate" is what physicians use to estimate the outcome of an illness. For cancer it's usually a "relative" rate: that is, it's the percentage of patients alive five years after diagnosis divided by the survival rate of a corresponding age/sex group that don't have cancer.

Finland has been particularly dramatic in this context. Up to the 1970s, it had the world's highest death rate from cardiovascular disease and was also high in the cancer league (mainly owing to the factors we'll come back to later, such as heavy tobacco use, high-fat diet, and low vegetable consumption). The effect of a national effort to improve these aspects of lifestyle has been to reduce heart disease in men by at least 65 percent and to reduce cancer mortality to one of the lowest rates in the world, thereby extending the average life expectancy of its citizens by over six years.

So there is some good news but to the sobering stats we must add the conclusion of the International Union Against Cancer that the number of new cases each year will rise from the current 12 million to more than 16 million by 2020. Many factors will contribute to this rise, notably the drift of people in developing countries from living in the countryside to living in towns with the associated exposure to pollution, to say nothing of junk food. Then there's smoking, set to increase worldwide, and viral infection, which is responsible for much of the cervical and liver cancer burden. Even so, the most dominant factor is that most of us are living longer—so we have more time to develop the disease.

Increased longevity is such an important factor in rising cancer rates that it's worth having an "ageist interlude" before we hit causes, but first let's just put cancer in its statistical place. The 7 million it kills each year—13 percent of the annual 56 million deaths in the world from all causes—comes second to mankind's biggest scourge, heart disease, which claims over 13.5 million deaths per year (24 percent). That's the total from what doctors would call ischemic heart disease, cerebrovascular disease, and hypertension. The next biggest killers all fall some way behind cancer, namely, HIV/AIDS (4.9 percent), tuberculosis (2.7 percent) and malaria (2.2 percent). Much less well known is the figure of 2.2 million children (3.9 percent) who will die from diarrhea this year, largely caused by picking up bugs through living in insanitary conditions.

A MATTER OF AGE

The longer we stick around, the more likely we are to develop cancer, despite the image of longevity morphing into an idyllic lifestyle for those in their "golden years." First of all, what we define as those golden years varies from culture to culture; according to the US Census Bureau, in Andorra you'll live to be nearly 84 (on average, that is), whereas if you're a male living in some parts of Glasgow 54 is the cut-off age, and in Swaziland the average life expectancy is under 32 years. So you would guess, therefore, that not many Swazis die of cancer, and you'd be right: fewer than 5 percent compared with about 30 percent of UK denizens. You might also guess that the spectrum of cancers they do get is a bit odd and, indeed, in countries where the AIDS epidemic is out of control (such as Swaziland) Kaposi's sarcoma is the most common cancer.

Both heart disease and cancers are diseases of old age. Men over the age of 50 who have never smoked are more likely to die of heart disease than anything else. More than two thirds of cancers first show themselves in people who are more than 65 years old. However, the average age of populations is steadily rising both in the developed world and in developing countries. In the Bronze Age the average lifespan was 18 years; the average worldwide is now 66—and that has risen from about 35 in the early part of the twentieth century. So, of course, the cancer and heart disease figures are rising and the grim prediction for 2020 is perhaps not so surprising. By then there will be 30 million individuals with the disease, 70 percent of them in the developing world. Of course, not all cancers wait until we are using walkers (Zimmer frames in the UK) before they appear. The entire category of childhood cancers, by definition, appears in the first 14 or so years of life. In addition, most adults will know of someone who has succumbed to a particularly aggressive cancer in early life. So cancers do strike the young, but these are mercifully rare and, from an overall statistical point of view, cancers are indeed something to look forward to in old age.

In general, therefore, the longer you live, the more likely it is that a cancer will appear somewhere in your body, but at least that means two out of every three only show themselves after the age of sixty. If you take the mortality figures and draw a graph showing the number of deaths and the age at which they occur, you get a straight line. And, somewhat surprisingly, that's true for pretty well every major cancer, regardless of country, sex, or race. The straight-line relationships between death rate and age reflect the fact that cancers are mostly caused by a build-up of specific events—on average, one occurs every ten years—and typically you need about six of them for a cancer to emerge. This hugely important deduction was made sixty years ago by the Finnish-born Carl Nordling—important because it was the first evidence that a cancer cell contains a number of mutations that are picked up over a long time. This tells us that although, as we have seen, the major cancer types can vary widely around the world, their development is underpinned by similar, cumulative molecular events.

From this brief look at numbers, we can therefore say two main things. Cancers are driven by collecting a small group of mutations and, to a considerable extent, they are self-inflicted. That is to say, mutations can be caused by a wide range of things that affect our bodies but we have some influence over these—we have a degree of choice in the way we lead our lives, and what happens to us is a consequence of our lifestyle, where we live and how we live. The problem with "lifestyle" is that sometimes it's a bit tricky to define what is a good way to live and what is not.

TAKING A LOOK AT OUR (MOSTLY) INNER SELVES

Before we go on to causes, for those of us fortunate enough not to be anatomists, a quick body guide might be useful. It's true that cancers can affect parts that are readily touchable (not now, please) but in general they go for submerged bits. The small intestine and the large intestine together make up the bowel. Cancers occurring in the colon, rectum, cecum, or appendix are therefore colorectal cancers, sometimes called colon cancer or just bowel cancer. The principal genital cancers affect the cervix, ovary, and prostate: if you want a more detailed map of those bits, Cancer Research UK know their way around (http://www.cancerhelp.org.uk/).

COMPARING CAUSES

History (and even epidemiology and anatomy) are fine things, but we need to get on to the picture of cancer right now by looking first at the main causes: what you might call the Beastly Biologicals. You know what's coming—ultraviolent radiation, abhorrent alcohol, truly terrible tobacco, wretched radon, mutating mobiles. The questions are: How bad are they, and what can we do about them?

Radiation

In our voyage through the history of cancer, one name we didn't meet was Heinrich Rudolf Hertz. "Of course we didn't," you chorus, "he was the German guy who showed that light comes in waves—nothing to do with cancer." So how come everyone knows about Hertz? Answer: his work led to the invention of radio and television and the unit of frequency that was named after him.

Everything that we see in the world around us comes, quite literally, in the form of Hertz's waves, or electromagnetic radiation, most easily thought of as a stream of particles called photons that carry energy at the speed of light. If that brings to mind flowing water as a stream of H_2O molecules, that's fine because radiation does behave like a wave even in that it too is made up of particles. The energy carried by a photon depends on its frequency and the term "electromagnetic spectrum" means the vast range of frequencies that gives rise to all forms of radiation. Frequency means the number of cycles per unit time—one wave cycle per second being one Hertz. The higher the frequency, the greater the energy: ultra-violet (UV) radiation and X-rays have higher frequency and more energy than visible light.

Our eyes can detect only a tiny region of the electromagnetic spectrum—the visible spectrum. Slightly higher frequencies take us into the UV range where the energy can now damage living matter directly. Other well-known regions include radio waves in the low-frequency range and X-rays in a higher-frequency band, which gives them enough energy to pass through solid matter, including the human body.

Go to higher frequencies still, and you enter the world of "ionizing radiation," which includes gamma rays, produced by subatomic particle interactions (e.g., in radioactive decay), and X-rays that result from high-speed electrons colliding with metal. Alpha and beta particles are also forms of ionizing radiation released when unstable isotopes undergo radioactive decay. Alpha particles readily give up their energy—they're absorbed by paper and can't pass through the dead outer layer of our skin—and are only a hazard if inhaled. Beta particles are smaller than alpha particles and some can penetrate the outer layer of skin. Gamma rays and X-rays are packets of electromagnetic radiation that can pass through all but the densest of materials and travel a long way through air—hence lead and concrete are used as shields.

Radioactivity is measured in becquerels (Bqs), named after Henri Becquerel, one Bq meaning that one nucleus decays per second. For a long time, radioactivity was measured in curies (Ci), named after Pierre and Marie Curie but, as 1 Ci = 3.7×10^{10} Bq (that is, an awful lot), the Bq is a handier unit. All three were French and shared the 1903 Nobel Prize in Physics but, as the Curies will probably always be remembered as the first, and so far only, married couple to win a Nobel Prize, perhaps it's fair enough for the name to go to Henri.

Ionizing Radiation

Ionizing radiation, and indeed UV radiation, are important because, as we've noted, they can damage living tissues, most significantly by causing mutations in DNA that can lead to cancer, as we shall discuss later. It is therefore a good idea to avoid them as much as possible by not swimming in the effluent from nuclear installations and keeping well away from atomic explosions. We should bear in mind, though, that we have evolved against a background of natural radioactivity that comes from radioactive elements in the earth and in rocks, trace amounts of which are also present in food and water, and in radiation from space. On average, about 87 percent of the ionizing radiation that hits humans comes in these unavoidable forms. The rest is artificial radiation of one sort or another, including X-rays used in medicine.

By way of reassurance before confronting the more extreme forms of radiation, it's instructive to spend a moment considering potassium. That may seem an odd idea because potassium is not exactly a high-profile chemical: everyone knows sodium is important, not least because of the use of saline drips—prominent in TV hospital dramas—to replace fluid in the circulation of accident victims. That's because sodium is one of the main constituents of blood: we have about 3.4 grams of it in every liter, only a little bit less than the normal level of glucose. There's about ten times less potassium in blood but, inside cells, the balance is almost exactly reversed—much more potassium than sodium. We'll look at what makes a cell in Chapter 4, but for the moment, the point to remember is that each cell is surrounded by a membrane, and one of this membrane's jobs is to keep at bay huge levels of sodium and potassium that would like nothing better than to even themselves out. Membranes therefore are two-faced dams, an occupation that's very expensive in terms of the energy needed to keep pumping sodium and potassium to sustain these concentration gradients—and you can be certain that they wouldn't put in so much effort if maintaining that balance wasn't essential.

So potassium as well as sodium is jolly important to the efficient functioning of the human body. The bad news is that it's the major radioactive emitter in our bodies. There are three forms of potassium that occur naturally (different forms of an element are called isotopes): two of these are stable, but potassium-40 is radioactive and, what's more, its half-life is 1.3 billion years (half-life is how long it takes for radioactivity to decay to half the original amount)—so it's a good job that potassium-40 makes up only 0.012 percent of our potassium. We can't avoid it because it's in the earth, though we increase the amount we eat by using fertilizers that contain nitrogen, phosphorus, and potassium, which add several thousand curies a year to US soil. This means that the fruit and vegetables considered essential in a good anti-cancer diet contain lots of potassium including its radioactive bit. Because cows insist on eating grass, if you drink milk, you will be downing about 74 Bq in every liter.

Ingesting chemically fertilized food, however, is not the only way we expose ourselves to greater amounts of potassium-40. When you sleep with someone, you expose yourself to this form of radiation. Your paramour's potassium-40 is decaying all the time and, nuclear physics being such a precise science, you can work out that, if your partner weighs 70 kg, 4,400 nuclei of their potassium-40 decay every second (they're emitting beta particles and gamma rays). That isn't a bid to promote potassium radiation as a contraceptive by putting you off sex, it's merely to point out that we have evolved in the presence of ionizing radiation from all directions and, in terms of propagating the species, you'd have to say we've not done a bad job of handling its perils, of which potassium-40 is one of the less well known.

Passing from those forms of radiation we can't avoid to those we might decide it's more useful not to, we should consider briefly exposures from radiation used in medicine. This mainly means X-rays and computed tomography (CT). In a conventional X-ray the radiation is directed at a part of the body and what passes straight through is collected on a photographic film or digital sensor, giving a two-dimensional image of the sort familiar to anyone who has ever broken a bone. CT also uses X-rays to acquire two-dimensional images but, from a large number of such images taken as the radiation beam moves through the body, a three-dimensional picture can be pieced together. This can give a picture of whole organs, and it has become an immensely powerful diagnostic tool since its introduction in the early 1970s. For inventing it Godfrey Hounsfield and Allan McLeod Cormack shared the Nobel Prize in Physiology or Medicine in 1979.

The biological damage caused by radiation is measured in units called the sievert, named after the Swede Rolf Sievert. The actual dose of radiation absorbed is measured in units called grays, but the sievert represents an attempt to take into account other factors (type of radiation, exposure time, body part exposed, and the volume thereof) to come up with a measure of "relative biological effectiveness." A typical chest X-ray requires a dose of about 0.04 millisievert (mSv); for a corresponding CT scan, the figure is 8 mSv. As our annual dose of "unavoidable" natural radiation is about 3 mSv it's probably safe to say that these typical medical exposures are not a serious hazard—so long as you aren't a baby in the womb.

On the other hand, radiation is, of course, one form of cancer therapy, the strategy being to use its lethal capacity by targeting tumor cells. For this, much higher exposures are required, which has driven the development of increasingly sophisticated machines that can target precisely the contours of the tumor.

Radon

In most parts of the world the largest source of ionizing radiation is radon, a product of radium. Radon is an inert gas that has a half-life of four days and arises from radioactive decay of uranium-238, which is present throughout the Earth's crust. Worldwide, one million deaths are caused by radon, with smokers being more vulnerable. Significant radon concentrations normally accumulate only inside buildings, but this can readily be prevented by installing a sealed membrane at ground level. Although there is a statutory level at which this preventative measure must be applied (200 Bq/m_3), it's been estimated that 85 percent of deaths caused by radon arise from lower levels for which preventative measures are not legally required.

Ultraviolet Radiation

The energy carried by UV radiation is sufficient to break the strongest chemical bonds (covalent bonds) and thus damage biological molecules. It's therefore a good job that almost all the radiation reaching the Earth from the Sun is absorbed by the ozone layer. The small amount of UV that does get through is important because it is responsible for producing vitamin D in the skin, a deficiency of which causes a high proportion of premature deaths and has been linked to many diseases including cancers.

UV radiation is both a carcinogen and necessary for good health, so its effects become a matter of balance. To facilitate the balance, we make melanin pigments in our skin that absorb UV. The cells that do this are melanocytes (they're also present in the eye and the bowel). They make both black eumelanin and a second type of pigment that's red, phaeomelanin. When we get suntanned, it's because our skin makes eumelanin as a protection against UV light. Dark-skinned people make more eumelanin and are thus better protected against skin cancer. The evolution of lighter skin in the peoples of northern climes may have been a response to lower levels of sunshine, thereby boosting their vitamin D production. Consistent with this notion is the relatively high incidence of rickets (one of the most frequent childhood diseases in many developing countries as a result of malnutrition) in more recent Indo-Asian immigrants to the UK, particularly vegetarians. The predominant cause is vitamin D deficiency: who would have thought it—these guys just don't get enough sun in the UK!

Melanocytes have come to prominence in the cancer field in recent years because their uncontrolled growth gives rise to a malignant tumor called a melanoma (Plate 5). Melanoma is a relatively rare form of skin cancer—the non-melanoma skin cancers are the most common—but it is the most serious. It causes three-quarters of all skin cancer deaths, and each year there are over 68,000 new malignant melanoma cases and 8,700 deaths in the US, 8,000 new cases and 1,800 deaths in the UK, and worldwide over 160,000 new cases and 48,000 deaths. In Britain and the US it is the second most common cancer in young people (aged 15–34) and the incidence is rising by 1 percent per year in both countries. As you might suspect, due to the amount of pigment in the skin, white Americans are almost thirty times more likely to develop melanoma than are African Americans. For non-melanoma skin cancers, UV exposure appears to be a major cause and, having once had this condition, increases the risk of future malignant melanoma. Because malignant melanomas develop from moles on our skin, they are the easiest cancers to detect at an early stage; if you have any such mark on your skin that changes color, size, or shape, you should seek medical advice. The encouragement to do so is that, if identified early, melanomas can be treated by surgery alone and the prognosis is then excellent (with a five-year survival of over 95 percent).

Low-frequency Magnetic Fields

In most countries the electrical power systems run at a frequency of 50 Hz and at about 230 volts (in North America it's 60 Hz and 120 volts). This means that most of us, whenever we are indoors, are surrounded by a network of wires carrying alternating currents producing corresponding electric and magnetic fields. You may escape by going for a walk in the country, but bear in mind that the overhead transmission lines that distribute electricity operate at several hundred kilovolts and they pop up all over the place, as do sub-stations, overhead electrified railways, etc., all radiating electromagnetic fields (EMFs).

Over the last twenty-five years a question that has frequently surfaced in the media is whether exposure to these EMFs can contribute to the development of cancers. The emotional temperature of the ensuing debate has undoubtedly been raised by the fact that EMFs have been incriminated in childhood leukemia. This is a very rare condition—there are about 500 new cases a year in the UK and about 2,200 in the US—but it is, of course, very emotive. The furor has sparked many epidemiological studies and a huge number of experimental projects that have covered a wide range of cancers. It has, however, proved to be a very intractable problem, mainly because the magnetic fields we are bathed in by these sources are extremely weak and there is no established mechanism by which they can affect our bodies. The understandable level of public concern has led to the establishment of independent bodies both in the US (EMF RAPID Program) and the UK (The EMF Trust) to support high-quality research into EMF effects. Thus far, the upshot of these efforts is that there is no convincing (i.e., consistently reproducible) evidence that EMFs could cause cancer.

High-frequency Magnetic Fields: Mobile Phones

The major concern over mobile (cell) phones has been the suggestion that their use increases the risk of two types of brain tumor (called acoustic neuroma and glioma) and that youngsters are particularly susceptible. As with childhood leukemia, a major epidemiological problem is that these cancers are rare in adults (one in 100,000 for acoustic neuroma and one in 30,000 for glioma) and even rarer in children. It is therefore difficult to show that using a mobile phone makes cancer a bit more likely. Mobiles work in the ultra-high frequency (UHF) radio frequency (RF) range (300–3000 MHz) where the energy is a tiny fraction of what's needed to break the weakest chemical bonds, the most likely way of producing a biological effect. As with power lines, therefore, there's no evident way for this sort of radiation to cause cancer and it's very difficult to design experiments to detect effects. Given these problems, the main approach has been to ask whether mobile users get cancer more often than non-users. This isn't easy either as you need tens of thousands in both groups to get statistically solid results. Nevertheless, there have now been several studies and, while the results haven't been absolutely clear-cut (they never are, unsurprisingly, when you ask people about their behavior—in this case how often they use their mobile), the upshot has been that "the current balance of evidence does not show health problems caused by using mobile phones," to quote the UK Department of Health. They do add, however, that children be "discouraged" from making "non-essential" calls while adults should "keep calls short."

We might leave the last word on the risks of mobiles to the exhaustive UK Government-commissioned Stewart report to the effect that mobile phones do present one major risk to health, namely when they're in the hands of motor vehicle drivers. This risk appears, interestingly, to be undiminished by the use of "hands-free" sets.

Tobacco

These days, one's navigation system must have taken you to planet Zog for you to be unaware that avoiding tobacco smoke is even more important than sensible eating in keeping cancer at bay. The estimate for the US is that 30 percent of cancers are caused by smoking, with poor diet accounting for 25 percent. The World Health Organization (WHO) has estimated that in the twentieth century tobacco-associated diseases killed 100 million people; in addition to cancers, these diseases included chronic lung disease and cardiovascular diseases. The attraction of smoking appears to be that it is both relaxing and stimulating (how does that work?) and that it improves our "image." The 100 million figure is thus a provocative reflection of what man can do when trying to better his lot, especially if you compare it with some of the major efforts at human destruction that characterized the last century, the Second World War or the rule of Mao-Zedong, for example, generally credited with about 55 million and 70 million dead, respectively.

The key study that conclusively linked smoking to lung cancer was published in 1950 by Richard Doll and Austin Bradford Hill (Plate 6). That and subsequent reports certainly influenced the attitude of the public to smoking, and in the ensuing fifty years the number of UK smokers has gradually declined by about half although, as the sketch shows, it took about twenty years for any effect to kick in. It's worth emphasizing that the smoking data are about as good a demonstration of cause and effect as you could imagine. It is true that a number of pollutants can give you lung cancer but, as near as makes no difference, before we smoked we didn't get the disease; as the number of smokers went up, so did lung cancer, and it's only since we've stopped lighting up that the death rate has started to fall. In a long-overdue response, the UK

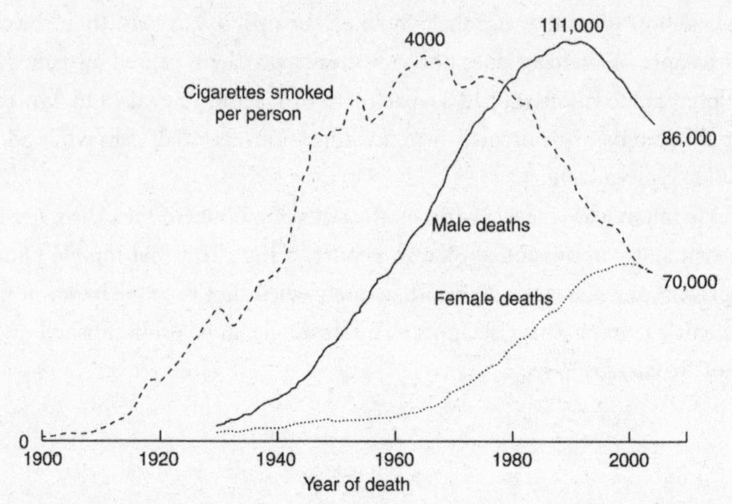

The long climb downhill: The gradual decline in UK lung cancer deaths since 1970. *The disease remains essentially untreatable, and the drop from 111,000 to 86,000 in the number of male deaths is almost entirely due to a decrease in the number of smokers. The pattern in women is similar but lags behind that of men and has only begun to decline since 2000. The US trends are similar.*

government has finally promoted an anti-smoking campaign, which has certainly made life more pleasant for non-smokers venturing into public places. Consistent with these developments, UK national statistics show that for those over sixteen years of age, the prevalence of smoking fell from 28 percent in 1998 to 24 percent in 2002. So we're going in the right direction but, as yet, this is hardly a startling drop, and lung cancer still kills 33,000 per year—about 100 people a day. Cancer Research UK tells us that rather more than that number (450, in fact) of children take up smoking *every day*, which perhaps explains why, of the almost 7 million between the ages of 16 and 24 in the UK, over 2 million smoke. Almost as many drink more than twice the recommended daily alcohol limit at least twice a week, and one million have used an illegal drug within the last month; in terms of the national effort to improve our health by reducing smoking and generally steering our young in the right direction, a mild summary might be that we could do better.

Perhaps the most depressing aspect of the smoking saga is that the first statistical evidence linking lung cancer and cigarette smoking was published not by Doll and Hill in 1950 but eighty years ago in the 1920s. By 1935 the German physician Fritz Lickint felt able to write that there was "no longer any doubt that tobacco played a significant role in the rise in bronchial cancer" and to coin the term "passive smoking." Because those pre-war studies were, however, carried out in Germany and published in German, they have tended to be ignored.

The gloom provided by the WHO figures for twentieth-century deaths due to smoking is as nothing compared to their predictions for the current century. You may wish to sit down with a stiff drink before reviewing these. The figure of 5.4 million a year that tobacco use kills now (that's one every six seconds) will rise to over 8 million (a year) by 2030. That sounds slightly at odds with the gradual decline in smoking in the Western world but, of course, the problem lies everywhere else. Or to be more precise, the problem lies largely in the tobacco companies of the Western world, and the developing world will pay the price. Of the more than one billion smokers, over 80 percent live in low- and middle-income countries. Currently there are 200,000 tobacco-related deaths annually in Africa, and it is there and in other underdeveloped regions that cigarette smoking is being heavily promoted by the manufacturers. It is the people of these regions who will, if this trend continues, contribute 80 percent of the 8 million dead. China, on the other hand, now rolls its own, so to speak, by making over 2,000 billion cigarettes a year, which goes a fair way to accounting for their half-million lung cancer deaths and the fact that by 2030 well over 3 million Chinese will kill themselves by smoking—3 million *every year*, that is.

All this is to say nothing of the children; while all this smoking is going on, children, who as ever have no control over what adults do, are breathing the smoke-polluted air (about half of all children are impacted by second-hand smoke).

The problem of involuntary (passive) inhalation, recognized so many years ago by Lickint, has been confirmed by numerous studies and has prompted legislation in many countries banning smoking in public places. In the US things are a bit more idiosyncratic, with some states having banned smoking in all enclosed public places but others having no state-wide prohibition. The most convincing evidence for second-hand effects comes from long-term studies of non-smokers living with smokers, for which pleasure their lung cancer risk goes up by 20–30 percent; similar figures have been found for exposure to cigarette smoke at work. As you would expect, the risk increases with time: twenty-one years of smoky domestic union and the risk of lung cancer in the non-smoker is increased by 25 percent. It's possible to measure exposure to tobacco smoke from the blood levels of a substance called cotinine, which is produced from nicotine. Somewhat alarmingly, a recent US survey found much higher levels of cotinine in children living in apartments than in those living in detached homes. The inference is that a substantial number of children are at risk from smoke diffusing into their homes from their neighbors—and that's in addition to more than 30 percent of American children who are currently exposed to second-hand smoking in their own homes and cars. The problem is further complicated by the dawning realization that "third-hand smoke"—that is, residual air-borne contamination present long after cigarettes have been extinguished—may also be a significant threat. It should also be borne in mind that passive smoking contributes to a range of other diseases, most notably asthma in the children of parents who smoke.

The major target for tobacco-promoted cancer is, of course, the lung, and smokers are about thirty times more likely to get lung cancer than non-smokers. But the effects of smoking are not confined to the lung: almost all the major cancer types are between two and six times more likely to develop in smokers than in non-smokers (i.e., mouth and throat, bladder, esophagus, pancreas, stomach, liver, cervix, kidney, and some leukemias).

It is perhaps surprising that most studies have found no link between smoking and breast cancer, or indeed prostate cancer or endometrial cancer of the uterus. Surprising because chemicals in tobacco smoke have been shown to cause breast cancer in rodents, and such chemicals have also been detected in both breast tissue and breast milk. The most persuasive of such studies compared more than 58,000 women with the disease with 95,000 who were disease-free, groups large enough to separate effects of smoking from those of alcohol. However, there are other surveys that suggest there may be a link. The California Environmental Protection Agency concluded in 2005 that, in women who were mainly premenopausal, passive smoking could be associated with breast cancer, and the US Surgeon General's 2006 report described the evidence as "suggestive but not sufficient." The age at which women start smoking does appear to be very significant in that smoking within five years of their first menstrual cycle almost doubles the risk of developing breast cancer before they reach menopause. This may be because teenage breast tissue that is still developing is more sensitive to smoke carcinogens.

One reason for this unclear picture may be that smoking reduces the level of estrogens, but this level is high in premenopausal women so that smoking may exert a major effect only in postmenopausal women.

It seems too obvious to need saying that it is best never to start smoking. But, if you have succumbed, all is not lost! Yet another major survey conducted by Richard Doll and his colleagues has shown that giving up (smoking rather than life, that is) will improve your chances of avoiding lung cancer. The earlier the better, of course, but even those who cease after the age of fifty reduce their risk by over 60 percent. As Richard Peto, the epidemiologist and long-term collaborator of Richard Doll, puts it, "smoking kills, stopping works."

Finally, while it is certainly true that governments have been appallingly weak in tackling this problem, progress was also obstructed by cigarette manufacturers preventing access to their own evidence of the lethal properties of cigarette smoke. It is still not as well known as it should be that there is definitive evidence that specific chemicals in nicotine can cause mutations in key genes that protect us from cancer. That's about as rock-solid as scientific evidence can get.

It's probably obvious that lung cancer and other consequences of smoking are not simply a matter of how much and for how long you indulge in the weed. As with more or less everything else, our genes play a role. There is now persuasive evidence that minor variants in a number of genes affect all stages of smoking: how likely we are to start, how much we smoke, and how easy it is to give up.

A last word from the lungs, so to speak, about asbestos, which continues to kill about 90,000 a year worldwide. Its nastiest effect comes from the damage the fibers do to the delicate sheets of membrane that cover internal organs. The lung is especially vulnerable, the cancer caused in this case being mesothelioma. Here the initial cause isn't mutation but inflammation that, if sustained, can lead to DNA damage, a point we'll come back to. For asbestos the effect is indeed "chronic." The time lag between exposure and evident disease is typically between twenty and forty years, which is why the death toll remains high even in countries where the law now reflects its hazards—over 2,000 in the UK and nearer 3,000 in the US annually.

Alcohol

Mankind has been brewing alcohol since at least the Stone Age by fermenting sugars from fruits or cereals to give ethyl alcohol (commonly called alcohol) with the release of carbon dioxide. The problem is not the making but the drinking, when 20 percent is absorbed by the stomach and the small intestine from which it passes into the bloodstream en route to all tissues and organs in the body (though it's not taken up by adipose tissue because it doesn't dissolve in fat). Eventually, most gets broken down in the liver by enzymes: alcohol dehydrogenase (ADH, which makes acetaldehyde) and aldehyde dehydrogenase (which makes acetic acid, aka ethanoic acid or vinegar). Acetic acid can be converted to carbon dioxide and water or used to make fatty acids. A small amount of our alcohol intake (about 5 percent) passes through the kidney to be excreted in urine. A similar proportion is breathed out via our lungs, which is why the breathalyzer test works: the gizmo uses potassium dichromate to oxidize alcohol to acetic acid. When this happens, the red potassium dichromate is converted to chromium sulfate, which is green, and it's the extent of the color change that tells the nice policeman how far you've transgressed.

It must be true that just about everyone over the age of five knows what the most obvious and immediate effects of drinking alcohol are. Incidentally, five is the age when alcohol can legally be consumed in UK homes; France and most American states don't have even that restriction although, of course, in some countries it's illegal at any age! Oddly, however, and a bit like one or two other utterly familiar things that appear to be very simple but are in fact amazingly complex—soap bubbles come to mind—how alcohol really works is still not absolutely clear. Alcohols are, as we know, very soluble in water so they are carried round in the blood and interact with all cells, and chronic exposure to alcohol changes the way cells work, especially those of the liver, which is why cirrhosis is a major long-term effect of drinking. Its immediate effects, of course, arise because it goes straight to our heads—alcohols cross the blood-brain barrier that enables them to reach brain cells and so affect them directly. We should also record the inescapable fact that women, being on average smaller than men, are more affected by the same alcoholic intake. The difference is an effect of concentration (i.e., dilution) though the result may be a loss of concentration of the mental variety, if nothing else. Things are made worse by the fact that nature has dealt women a rather unhelpful hand. We saw that ADH breaks down alcohol, but it starts to do that even before the alcohol leaves the stomach. Remarkably, humans have seven different kinds of ADH. Nevertheless, women, especially young women, make less ADH than men, so that means proportionally more alcohol gets out of their stomachs and into the bloodstream.

Circulating alcohol can affect all tissues but in particular it heightens the effect of signaling molecules in the brain. We'll come to cells and how they receive signals in chapter 4, but for the moment the reason alcohol affects our brains is that it's really a local anesthetic (ether, one of the first anesthetics used in medicine, is, in effect, two molecules of alcohol joined together). This is why you are told not to drink alcohol in the two days prior to having a general anesthetic.

Regardless of how it works, alcohol consumption is clearly linked to an increased risk of developing breast cancer. In developed countries about 4 percent of breast cancers are attributable to alcohol, the risk increasing with the amount consumed. One plausible explanation of its effects on the breast is that it makes cells produce more estrogen, which in turn increases cell growth.

Compared with non-drinkers, women who consume one alcoholic drink a day have a very small increase in risk. Those who have two to five drinks daily have about 1½ times the risk of women who drink no alcohol. The American Cancer Society recommends that women limit their consumption of alcohol to no more than one drink per day, advice worth following as excessive alcohol use is also known to increase the risk of cancer of the mouth, throat, esophagus, bowel, and liver.

Time for Tea (and Coffee)

As well as being major risk factors for lung cancer and breast cancer, tobacco and alcohol also increase the risk of head and neck cancers, a term that collectively refers to cancers of the lip, mouth, nasal cavity, pharynx, and larynx. This leads us to a final word about drinking. We think of the word as referring to alcohol but of course we drug ourselves with other fluids. Tea and coffee are the most common hot drinks in the world, and it's well known that they contain caffeine, which can cross the blood-brain barrier and act on the central nervous system—this means that the everyday tea and coffee in our pantry contain a psychoactive stimulant drug. Caffeine is also a well-known diuretic—i.e., it increases the volume of urine produced—though regular users get used to this side effect. Both the coffee plant and the tea bush contain caffeine and in fact there's more caffeine in tea than coffee by dry weight but you get much more in a cup of coffee than one of tea. So one might well ask, with all this drug abuse of our throats, is there any evidence that drinking tea or coffee gives us cancer? The most comprehensive analysis so far (of results from nine studies) shows, perhaps slightly surprisingly, that there is an inverse association between drinking caffeinated coffee and the risk of cancers of the mouth and pharynx. In other words drinking coffee protects against these cancers, and the more you drink, at least up to four cups per day, the greater the protection. The picture for decaffeinated coffee is less clear but at least it does not appear to increase risk. Tea drinking showed no association with head and neck cancer. The leaves of the *Camellia* tea plant are a particularly rich source of antioxidants, and on a per-serving basis coffee provides even more of them. Substantial proportions appear to be absorbed by the body, so it is possible that they may protect DNA from mutation.

Infection

Having earlier made the passing acquaintance of diarrhea, we should note that living conditions, and especially sanitation, bear on cancer partly because chronic infection usually reduces the efficiency of the immune system, which is the defense the body turns on when we get infected with germs. It has gradually emerged that this system also recognizes tumor cells and can attack them too. Quite how it does this is still not completely clear, but it is unquestionably one of our anti-cancer defenses, and its impairment releases one of the brakes on tumor development, a subject we'll come back to in Chapter 7. Although almost all cancers arise from genetic damage, some can be started by microbial infection, which means everything we think of as "germs"—bacteria, fungi, and viruses. This is most clearly established for stomach cancer in response to a bacterium, *Helicobacter pylori,* that gives rise to chronic infection and ulcers in the stomach. This bug can double the risk of getting gastric cancer and some more potent strains can increase the risk by 30-fold. About a quarter of us pick up *H. pylori* but the vast majority don't get stomach cancer, clearly indicating that other factors are involved. Another bug, the tuberculosis (TB) bacterium (*Mycobacterium tuberculosis*), not only kills 2 million people a year but is present in latent form in about one in three of us. It's been known for many years that patients with TB have an increased frequency of lung cancer, and recent experiments with mice have shown that chronic infection can indeed provoke malignant lung cancer.

There are, of course, numerous other infections that can strike us—one such, undeterred by his radio waves, killed Heinrich Hertz by setting off an immune reaction against his own blood vessels. The common theme that links infection by different germs, including viruses, to cancer is inflammation in the affected area; we'll look more closely at both inflammation and tumor viruses later.

Food

So far we've met a variety of things that may give cancer a helping hand, some more avoidable than others, some heavily incriminated, others where the jury is still out. But we've barely touched on the most obvious contributor of all: what we eat. There's no question diet is important because, like tobacco use, the numbers overwhelmingly show it's a major factor in worldwide cancer variation. The Western world always comes off worst in these comparisons, of course, with its junk food washed down with alcohol. Before we sum up the

THE UNCERTAINTY PRINCIPLE: WEIGHING THE EVIDENCE

There are really two ways to answer the question of whether what we do to ourselves, and particularly what we eat, affects our cancer risk. The first is what epidemiologists call a retrospective study: take a group of people who have (a specific) cancer and select a second group who are as similar as possible except that they don't have cancer. Then ask them what they have eaten for the last twenty or thirty years. These are sometimes referred to as observational studies and the behavior of each subject is self-selected—they ate what they fancied, which is good because the investigator hasn't influenced what happened. But you will already have spotted a bit of a pitfall: no matter how much statistical analysis you bring to bear, you can't make the two groups really comparable. Equally problematical is that the accuracy of answers to the question "What have you been eating?" probably exceeds only marginally that of the answers to surveys concluding that everyone engages in sexual congress 3.5 times per week.

The second type of study is prospective: this is an experiment, sometimes called a randomized controlled trial (RCT), in which the participants are divided into two or more groups and each group is given specific behavioral instructions (in this case, told what to eat) by the investigator and everyone waits to see what happens. This may take some time when dealing with cancer, but the main advantage of the method is that random distribution into the groups reduces bias. One of the biggest problems, of course, is that however hard they try, individuals will vary in how well they stick to the dietary rules. This method has been extensively used to find out whether agents in foods that have been thrown up, so to speak, by observational studies actually do confer protection against cancers.

Though easy to describe, such studies are complicated to design: you have to enroll an awful lot of people before the results mean anything, they take a long time and they are, therefore, very expensive. It might be added that, in the diet field, for the most part they have also been singularly uninformative, the clearest conclusion being that linking specific things we eat to cancer is a desperately tricky business. The guy who ate 25,000 Big Macs makes the point: the worst of his problem is probably not the polar bear's weight of fat he's eaten but the rather vague things in fruit and veg he's not had time to squeeze in.

evidence for the prosecution, however, it's instructive to look at how these associations are pinned down, even if pinning them down is a similar problem to the one we mentioned in the context of magnetic fields, where the major difficulty was not knowing what to measure as the cause. But for diet we know our causes—specific foodstuffs—and we know the end point—cancer. What can go wrong with an experiment designed around criteria we can pinpoint?

The Question of Diet and Cancer

A summary of one of the biggest prospective studies yet undertaken illustrates the scale needed for such projects. The Nurses' Health Study started in the US in 1976 to investigate risk factors for cancer and other diseases in women. Over 90,000 nurses have enrolled in this enterprise. Among its major findings so far are that the risk of bowel cancer is increased by eating a lot of red meat and reduced by taking folic acid in multivitamin supplements. Breast cancer incidence is unaffected by fat and fiber intake but is increased by one third in response to moderate amounts of alcohol (1–3 drinks per week), even though the latter makes you less likely to have a heart attack (unless, of course, you read the cancer predictions!).

Bowel cancer has been intensively studied in the context of diet partly because a fair bit of what we eat comes into direct contact with our intestines. Bowel cancer is also a well-defined, multistep process in which, without treatment, benign tumors develop into full carcinomas. In the early stages, polyps form on the wall of the bowel. These can be surgically removed and individuals so treated have been particularly studied to determine the effect of diet on recurrence. Oh, and each week in the US over 1,000 die from it—320 in the UK, 175 in Canada, 24 in New Zealand. Sorry to go on about it—just don't want anyone to think they can escape north of the 49th parallel or by taking pot luck with earthquakes. Three familiar dietary players that have come to the fore in the context of cancer—folate, calcium, and fiber—may, however, take a role in lessening the risk.

Folate

Folate (there are several naturally occurring forms) is a vitamin that helps to maintain DNA, so it's easy to see that if you were short of folate, you might be prone to cancer. A synthetic form of folate—folic acid—is used as a food supplement. A number of both retrospective and prospective studies suggest that folate has a protective effect, consistent with its importance in maintaining the integrity of our DNA. Furthermore, other studies have measured the levels of folate in blood and found an inverse association with bowel cancer—that is, the more folate you have, the better—consistent with the Nurses' Health Study. It's also worth bearing in mind that folic acid supplementation shows one of the best cause-and-effect examples in the 75 percent decrease in spina bifida that results from daily intake prior to conception.

So far, so good but, as usual in this field, there are conflicting reports, some of which conclude that diet supplementation can actually increase the risk. For folate you need enough to maintain your DNA in a healthy state, but if you have too much, it may block DNA repair, thereby helping to drive cancer. So, supplementing diet with folic acid might provide protection against bowel cancer for those with low levels of circulating folate but might be very unhelpful for individuals with higher natural levels. All of which indicates the dangers of advocating diet supplementation for general populations rather than investigating the need of each individual person and determining what will help to balance their unique, individual biochemistry.

Calcium

As with folate, there is a general view that plenty of calcium is good not only because it gives you strong bones but also because you are less likely to get bowel cancer. Again, there are prospective studies that support this view but, as ever, there are others, including one of 36,000 cases showing that taking more calcium has no effect. These findings scarcely make the case for calcium supplementation and, as with folate, may be confounded by individual variation—that is, additional calcium intake will not benefit those on healthy diets whose levels are normal.

Fiber

Dietary fiber, sometimes called roughage, is the stuff we eat but can't digest, and it does an important job in taking up water and generally helping our insides work. The well-publicized advice is that eating plenty of fiber helps to prevent bowel cancer, and there are many supporting surveys. Notable among these is the European Prospective Investigation into Cancer and Nutrition (EPIC) study showing that 35 grams per day of fiber reduced the risk by 40 percent compared with 15 grams per day. This is a particularly powerful contribution because it involved over half a million (520,000) people from ten European countries. Almost inevitably, there are other studies (e.g., the Polyp Prevention Trial and the Wheat Bran Fiber Trial) that show no protective effect and at least one concluding, somewhat improbably, that a relatively high fiber diet protects men better than women. In addition to the pitfalls mentioned earlier, a further cause of confusion may be the variation in study duration and the fact that follow-up periods are generally short relative to the many years over which cancers usually develop.

BALANCING THE ODDS

It is probably obvious that, when faced with a considerable number of independent studies about a specific factor with conclusions spanning all categories ("it's good, it's bad, or it's useless"), it might be worth putting them together and seeing if a more focused message emerges. This is what epidemiologists call a "meta-analysis," and it's really a way of increasing your sample size—hence the statistical power of the data. It's not absolutely straightforward because there's no point in including a study you think is so badly designed that it's worthless. So an element of judgment was required to combine data from thirteen separate studies to show that there was no effect of a high fiber diet on the risk of developing bowel cancer, a conclusion that was, of course, completely at odds with the EPIC finding.

So thinking about diet raises one of life's great problems—namely, that it often seems as though you can't win. Many foods contain DNA-damaging agents or something else nasty or are deficient in a specific essential nutrient. We're often encouraged to "eat our greens"—generally sound advice that would include cabbage, which has lots of vitamins C and K, and also to eat plenty of fiber, calcium, manganese, folate, and sundry other goodies. But cabbage also contains chemicals called goitrogens that can interfere with the function of the thyroid gland—which is why people who have thyroid problems are advised not to eat it. The cabbage dilemma might indeed make you think that maybe diet isn't quite such a simple matter after all. It is now nearly thirty years since Richard Doll and Richard Peto came up with the estimate that diet may contribute to one third of all cancers. Despite the fact that this figure is still widely accepted, there is very little direct evidence that links the action of specific chemicals in foods to the way in which cancers develop. Take the long-running antioxidant saga: the cumulative evidence suggests that, if anything, upping your β-carotene level by whiffling your way through stacks of carrots makes you more prone to cancer and heart failure rather than less. Take red meat, with or without carrots: of course it's good for you: we've evolved in part by eating it, it looks good, tastes good, and gives you lots of protein and iron. However, over the years a number of large and seemingly well-conducted studies have shown that you're a bit more likely to get bowel and stomach cancers if you eat lots of red or processed meat. Reviewing the epidemiology of food and cancer, the World Cancer Research Fund and the American Institute for Cancer Research recommend that for red meat we consume less than 500 grams per week and that "very little if any" of that should be processed—that is, smoked or otherwise cured. Bad news for me and for most of the population of Italy. Not great news either for the various bodies responsible for meat production who have been quick to point out that there are also large and seemingly well-conducted studies showing no link between meat eating and cancers. And you can't escape by becoming a trendy veggie because they too have an increased bowel cancer risk according to some studies.

The problem is that redness in meat comes from blood, specifically the iron-containing heme group in red blood cells that carries oxygen. Heme is broken down in our gut to give substances that introduce mutations into DNA. In addition, suppliers often add chemicals to meats to give color and flavor, and to stop bugs growing. These too can damage DNA, thus helping to promote cancer. The evidence that they can do this in rats is clear, but it comes with the rider that, generally, far higher doses than humans would ever consume are required to cause cancer, or that the chemicals cooperate with other tumor-causing chemicals to make them more effective—the rats will get tumors from the other (added) chemical: the meat-related chemicals just make them grow faster.

It's not just bowel and stomach that have been implicated as targets for meat-promoted cancers: there are several reports showing strong associations with lung cancer—and, of course, there are also several showing no association. You can readily see that, once you've got agents that can mutate DNA wandering around in the blood, lung cancer may not be far behind. A recent report of a nearly two-fold increase in lung cancer risk with meat eating is interesting because it notes many of the problems encountered in epidemiological studies. Probably the biggest of these is having to rely on questionnaires and interviews for the information about what folk ate last week, last year, over the last ten years. Quick: Do you remember what you had for lunch yesterday? You see the problem. It's worse than that, however, because meat is better eaten cooked than raw, and cooking at high temperatures produces DNA-damaging chemicals. How often you flip your burgers on the grill may even be important, so it's not surprising that there's a pretty bewildering range of results and recommendations out there.

THE COCHRANE COLLABORATION

You may by now have the impression that the only thing more tricky than carrying out clinical trials is making head or tail of the results. However, help is at hand in the shape of The Cochrane Collaboration, a nonprofit consortium dedicated to analyzing and summarizing the literature on healthcare interventions. It has literally been a lifesaver by providing doctors with comparative analyses of trials. Its other great virtue is that it is set up to inform non-scientists as well by giving summaries in "plain English." These explain, very briefly but clearly, why the question is being asked, how the trials were set up, the main results, and the conclusions. For dietary calcium one of the most recent reports concludes that, although there is evidence that calcium supplementation might make a modest contribution to the prevention of bowel cancer, there is not sufficient evidence to recommend its general use. A corresponding report on dietary fiber concludes that there is currently no evidence to suggest that increased intake will reduce the incidence or recurrence of bowel cancer within a two- to four-year period.

OBESITY

All this talk about eating leads almost inevitably to thinking about being overweight. Although a number of factors may contribute, the basic cause of this condition is metabolic imbalance. In metabolic terms our bodies expend energy in two ways, either through using it as chemical energy or as heat loss, so that expenditure is what we need to match in the energy we consume as food, no more and no less.

Over 300 million people in the world are obese. That is, of all the people in the world, more than one quarter are overweight (you could make 75 million normal-sized adults from the surplus). Body fatness is often assessed by working out the body mass index (BMI), which is calculated by dividing weight in kilograms by the square of height in meters. Recidivist Brits and Americans can do it in pounds and inches provided they remember to multiply the answer by 703 ($BMI = lbs \div (inches)^2 \times 703$). Whichever way you do the sum, a normal BMI emerges as from 18.5 to 24.9, 25.0 to 29.9 is overweight, and 30.0 and above is obese. The big problem, if one may put it that way, is that the number of people with a BMI over 30 is rising rapidly. The chances are 50/50 that a British reader will be overweight and more than 25 percent of them will be obese—three times what it was twenty years ago. In the US the proportion is 30 percent. The American Cancer Society has a spectacular series of maps of the United States coloring in red each state in which more than 55 percent of adults are overweight. In 1992 not one state was red—*not one.* Now turn to the 2007 map: this looks like *The Empire Returns* (British of course)—*every* state is red. *Every single one including Hawaii* and if you think the forty-ninth fell off their map and might help, think again: the overweight figure for Alaska is 66 percent. So, taken together, one could say that the Anglo-American axis could do better.

It turns out that a good many sporty types, e.g., rugby players, have BMIs indicating that they are obese. This is because BMI does not distinguish fat mass from lean mass, and it's not just rugby players that can be misled because different populations (European whites, Chinese, South Asians) show differing associations between BMI, body fat, and metabolism. So sometimes it's preferable to use waist circumference as a measure, which correlates better with metabolic risk (for European men and women over 94 cm and 80 cm, respectively, are the girths that define being obese: these may differ somewhat for other ethnic groups). Rather helpfully therefore, a very large study (on 359,387 participants in nine European countries over ten years) concluded that, indeed, simple waist measurement was the best index for risk of death due to being overweight. Ideally it seems we should be no more than 37 or 31 inches round our middle, depending on our sex (that's 94 and 80 cm). Any bigger and your risk goes up so that a man with a 47 inch waist is twice as likely to die from factors associated with being tubby.

Many people know that obesity can have a genetic cause and it's certainly true that it can be caused by mutations in some genes. Perhaps the best known of these is the one that allows us to make the hormone leptin, a critical regulator of energy balance and therefore of body weight. Mutations in the leptin gene that reduce the level of the hormone cause a constant desire to eat with the predictable consequence. But only a very small number of families have been found who carry leptin mutations and, although other mutations can drive carriers to overeating, they are even more rare.

So that leaves an awful lot of people who can't put the blame on the genetic hand they've been dealt and, although there are different ways of measuring the problem, we're stuck with facts that are every bit as overwhelming as those about cancer. Obesity currently kills 300,000 Americans each year. In the UK 1 in 11 deaths is linked to excess fat. About a quarter of our 11- to 15-year-olds are already obese, so the problem is going to get worse rapidly. The percentage of kids between 2 and 10 years of age who are obese is predicted to double in the twenty-year period from 1995—yes, under 10 years old!

Obesity is important because it dramatically increases chances of developing a wide range of life-threatening conditions including diabetes (in the US more than 1 person in 3 ends up with adult-onset diabetes), arteriosclerosis, hypertension, heart disease, age-related degenerative disease, sleep apnea, gallstones, and some cancers. The estimate is that if we all had a BMI of less than 25 there would be 12,000 fewer UK cancers per year. Obesity is a significant factor in promoting cancers of the bowel, kidney, liver, esophagus, pancreas, endometrium, and breast, and there is evidence that it may also contribute to gallbladder and ovarian cancers. Breast tissue is a mixture of different types of cells and its make-up determines what's called the density. This has got nothing to do with size or appearance: it refers to the fact that X-rays pass relatively easily through fatty regions but are absorbed by denser tissue, so fatty tissue gives dark areas in mammograms but "denser" regions absorb X-rays and are white. The balance is important because a high density carries a 4- to 6-fold increase in breast cancer risk. High tissue density also greatly increases the difficulty of detecting small tumors by mammography (Plate 7).

The risk of breast cancer is particularly significant in postmenopausal women. Before menopause the ovaries produce most of a woman's estrogen, and fat tissue produces the rest. After menopause (when the ovaries stop making estrogen) estrogen comes mainly from fat tissues. Having more fat tissue after menopause can increase estrogen levels and thereby increase the likelihood of developing breast cancer.

Unfortunately, it's a bit more complicated than that, as is almost always the case with the multi-factorial cancers. The risk is greater for women who put on weight as adults by comparison with those who have been overweight since childhood. Furthermore, there are differences between fat cells in different regions of the body so that waist fat appears to be worse than hip or thigh fat in terms of breast cancer risk.

While the upsurge in obesity is a relatively recent trend, there is evidence dating back 100 years for a link between the amount eaten and cancer. In rats and mice, dietary restriction, which means being fed between 10 percent and 50 percent fewer calories, reduces the incidence of at least some types of cancers and extends lifespan. Note that it's total calorie intake rather than the nature of the food that counts. Does this apply to humans? That's a bit more difficult to be sure about, but certainly some human tumor cells when grown in mice are very sensitive to dietary restriction. The explanation may be that when diet is restricted. the levels of key metabolic hormones decrease, particularly insulin. As well as being a major metabolic regulator, insulin is also a very potent promoter of cell growth—and potentially a driving force for cancer. Sustained low levels of such a factor may therefore be protective.

HOW CAN WE HELP OURSELVES?

Despite the huge amount of scientific experimentation and analysis behind this summary of cancer causes, the answer to the question "How can we reduce our risk?" is fairly simple. There are some things out there about which we can do nothing—radiation from the ground, for example—so forget them. You may feel that the radiation reaching us from the sun is also a bit beyond your control but remember the damage it can do, especially if you are fair-skinned, and protect yourself accordingly. There are others we can't completely avoid—the many foods that contain natural carcinogens (though you might watch out for the avoidable additives). But what you can do is match intake to output—in other words get off your rear end and do some exercise. If you guys need an incentive, how about the accumulating evidence that men in desk jobs are 30 percent more likely to get prostate cancer than those with more physical occupations, which is actually a bit perturbing when I consider how long I've spent parked on my keister writing about cancer.

In short, apply the rather old-fashioned approach of common sense. There's nothing naïve about this approach because it's a pleasure to report that, after analyzing all the epidemiology and scientific endeavor, it just about sums up the recommendations of the American Institute for Cancer Research and the World Cancer Research Fund. We should, therefore, pass on this good advice—as Oscar Wilde said, the only thing to do with it. He added, of course, that "it is never of any use to oneself."

1. Eat plenty of vegetables, fruits, lentils, beans, and whole grains such as brown rice and whole wheat pasta (at least two-thirds of any meal). These have a relatively low calorie content but their water and fiber content fill you up and, of course, they give you most of the vitamins and minerals you need.
2. Stay as lean as possible (without becoming underweight). A minimum of thirty minutes physical exercise a day will also help to prevent excessive body fat from forming, an established risk factor for cancers of the esophagus, endometrium, pancreas, kidney, bowel, and breast in postmenopausal women.

3. Don't eat too much red meat (no more than 500 grams cooked weight a week). Some red meat is good because some of the substances it contains are nutrients but others can damage the lining of the bowel.
4. Don't drink alcohol or anything with a high sugar content. If you can't face an alcohol ban (and it is regularly reported to protect us against heart disease), limit yourself to one (women) or two (men) drinks per day.
5. Don't eat too much salt.
6. Don't use food supplements unless specifically suggested to by your physician.

This very basic advice really makes the point that, despite the vast number of food and cancer studies, the difficulties in selecting study populations and carrying out such surveys mean that the results are rarely convincing. They also imply one other point that is worth stressing: with the exception of alcohol, removing anything from a balanced diet generally has a negative impact on health greater than any reduction in carcinogenic effects.

STRESS

Stress is another factor that may be associated with cancer. It's possible to put a number on stress, so to speak, by measuring the amount of cortisol in blood or saliva. Cortisol is a steroid hormone released from the adrenal gland in response to signals from the brain. Normally the level is highest early in the morning and falls during the day (diurnal variation), but it is increased by food, fasting, exercise, or stress. Its role is to provide energy when required by stimulating the breakdown of fat and proteins, raising blood sugar levels and blood pressure. On the other hand, cortisol can also indirectly increase appetite and promote fat deposition. Some merry souls at the University of Trier in Germany spent a fair bit of time working out what, in normal life, we find most "stressful" and came up with two things: making a speech and doing arithmetic in front of an audience. Five minutes of these is seemingly enough to push our salivary cortisol levels up 2- to 4-fold.

Studies of breast cancer patients have shown that about two-thirds may have abnormal cortisol profiles (higher, relatively constant, or maximal at abnormal times of day) and this group survive for significantly shorter times (3.2 versus 4.5 years) than those with normal cortisol profiles. The reason may be suppression of the immune system by raised cortisol levels because they have reduced numbers of white cells in their blood. A second strand of cortisol association is the evidence that night-shift work, which disrupts normal diurnal rhythms, is associated with increased incidence of breast cancer. This has been attributed to melatonin suppression but may well be due to perturbed cortisol variation. Despite all this, in the confusing way that cancer has, cortisol-type steroids can suppress the growth of some tumors and have been used as therapeutic drugs.

WHERE DO WE STAND? AND IS THERE ANY GOOD NEWS?

It has to be admitted that, at a quick glance, there doesn't seem to be much to be cheerful about. The predictions for the worldwide increase in cancer rates are almost beyond comprehension and, as we've seen in passing, there isn't anywhere to hide. Certainly not in the UK where in 2008 there were 310,000 new cancer cases, an increase of 54,000 compared with 1994. In particular, there was a 15 percent increase in breast cancer and sharp increases in bowel and prostate cancer. Of course, part of the reason for the increasing trends is that we are living longer and, as we saw earlier, cancers are mainly diseases of old age. Nevertheless, cancers also afflict the young. The 156,000 cancer deaths per year make up one quarter of all UK deaths, but that proportion rises to over a third for all deaths under the age of 65 and to one half of deaths in women under 65. This reflects subsets of cancer (e.g., childhood cancers and aggressive forms of breast cancer, as we noted earlier) that are relatively rare but have a big statistical impact on an age group that, by and large, doesn't tend to die from other causes.

Some encouragement may be gleaned from the progressively earlier detection of the disease together with the development of effective drug treatments. This has been particularly true for cervical and breast cancers and has meant that the developed world has seen a steady improvement in the five-year survival rate for breast cancer. Overall, in the richest countries about 50 percent of cancer patients survive the disease. However, in the developed world you are more than twice as likely to be diagnosed with cancer than elsewhere, with the result that in developing countries 80 percent of initial diagnoses are of late-stage, incurable tumors.

This shows that generally the world needs to improve screening programs, but we could also do much more by way of prevention. In developed countries only about 8 percent of cancers arise from infections, but elsewhere almost one quarter are caused by agents such as hepatitis B and C virus (liver cancer), papillomaviruses (cervical and ano-genital cancers), and bacteria. For example, 80 percent of deaths from cervical cancer are in developing countries. HBV and HPV vaccines are now available that are highly effective (chapter 9), and it is to be hoped that these will become sufficiently available to make a major impact on the incidence of cancers promoted by these viruses. Despite these wonderful contributions by science, it is difficult to be optimistic about controlling cancer in Africa, given the economic and logistical problems. The setting up in 2008 of AfrOx (Africa Oxford Cancer Foundation) to bring UK expertise to bear on the African cancer problem is at least one encouraging step. The most notable bacterial carcinogen is *H. pylori*, the gram-negative bacterium mentioned earlier as one of the causes of stomach cancer. There are drugs for treating this infection although, as most of those infected show no symptoms, we're clearly quite good at keeping it under control ourselves.

In the UK the picture is somewhat mixed. Better diagnosis and treatment have produced an annual decline in cancer deaths since 1983. The long-term survival rate is about 40 percent although the figure varies widely across the 200 or so distinct forms of cancer. Of these, as we noted earlier, four types (breast, bowel, prostate, and lung) account for half of adult cancer deaths in the UK. In the period from 1984 to 2004 the five-year percentage survival rates for the first three of these cancers improved, prostate from 42 to 74 percent, breast from 59 to 81 percent, and bowel from 40 to 50 percent. The most dramatic improvements of all have been for testicular cancer and some childhood cancers where the five-year survival rate has gone from being very low twenty years ago to over 90 percent. For other cancers, however, notably those of the lung and pancreas, there has been no significant change, and the rate remains unchanged at about 5 percent.

These large differences in the survival rates and in the impact of medical research arise from the fact that although, as we shall see, there are some basic features that apply to most if not all cancers, distinct molecular mechanisms drive the diseases, and they arise in different types of cell. For these reasons no single therapeutic regime is likely to be fully effective even against one cancer type, let alone against all. The one advantage of this diversity from the point of view of devising therapeutic strategies is that cancers present many potential targets for slowing or reversing their progress.

So there have been some advances in the UK but in terms of detection and treatment major problems remain that have more to do with the health system in Britain than with the biology of cancer. Despite the improvements mentioned above, UK five-year survival rates for common cancers are worse than the European average by 5 to 15 percent. Considering breast cancer, this gulf is even greater for deaths within six months of diagnosis. There are three possible explanations for this: (1) later diagnosis, (2) more aggressive forms of the disease in Britain, (3) lower standard of treatment. It seems likely that all three play a part and, tellingly, a 2001 review delicately concluded that "it is difficult to refute" the suggestion that breast cancer care in the UK has been non-uniform and sometimes inadequate. Comparison with the US for breast cancer tells a similar story that is mirrored by other cancers. Since 1990 the mortality rate has shown a steady decline in the UK, closely parallel to the decrease for white women in the US. However, the UK rate remains persistently about 20 percent higher—quite similar, in fact, to that for black Americans. Prostate cancer is something of an exception in that the death rate in the US is falling four times faster than it is in the UK, although the real difference may be less than this owing to differences in ways of attributing the cause of death; nevertheless, there is a significant difference.

Perhaps the most important point behind all the incidence and death figures is how well people do after diagnosis of cancer. In terms of five-year survival after diagnosis, the rates for all cancers in Europeans are significantly worse than in the US by about 47 percent versus 66 percent for men and 56 percent versus 63 percent for women, respectively.

Equally perturbing is the fact that where you live in the UK bears significantly on your cancer risk. The National Cancer Intelligence Centre has produced a "Cancer Atlas" that compares incidence and death rate from the twenty-one most common cancers in different counties of the UK. The differences reflect levels of smoking, drinking, poor diet, and social deprivation and show that regions of northern England and Scotland are cancer "hot spots." Their estimate is that if the worst areas could be converted to the best, there would be 25,000 fewer new cases and 17,000 fewer deaths a year: with about 157,000 cancer deaths per year, that would represent an 11 percent decrease.

One of the problems, of course, is that patients from poor backgrounds are more likely to be diagnosed with cancer at a later stage and thus adversely prejudice the efficacy of treatment. This is probably the reason for the disparity between the breast cancer death rates of blacks and whites in America. Even so, more subtle factors can also be involved. A report of 2001 (CancerBACUP) pointed out that, although most breast cancer treatment centers screened tumors for the presence of estrogen receptors, the methods and interpretation of the data varied so widely that a significant number of women were receiving unsuitable drug treatment. These problems are a strong argument for specialist treatment centers.

Finally, in looking for positive cancer developments, we should note the quite staggering amount of information about the molecular biology of cancer that has accumulated over the last twenty-five years. Given that it is a cancer biologist speaking, you might be tempted to question my reliability. It's nonetheless a fair point because, bewildering though the information avalanche has been—several "cancer genes" have been in the starring role of over 10,000 separate scientific papers—without an understanding of the underlying science at a molecular level, we cannot attempt rational approaches to cancer treatment and prevention. We will see shortly how this embryonic field, which unites knowledge of molecular interactions and structures with genetic engineering, has already had significant impacts in the clinical treatment of human cancers.

Throughout this book the emphasis is, of course, on scientific fact and method: how that has created our picture of cancer and informed our approaches to the treatment of these diseases. The perception of non-scientists generally seems to be that scientists lead mighty tedious lives that orbit exclusively around facts and methods. Put another way, they spend so much time peering down their microscopes that they lose sight of what life's about. Well, maybe the scientists you know are like that, but most of the ones I meet are a jolly eclectic bunch. What's more, far from becoming blasé about the life they study, they are much more inclined to periodic rapture about the wonderful things they're privileged to see. I sometimes try to explain this to students as a "wow!" moment and when I reveal that, on average, I have one every couple of months, the general response is, predictably, a skyward raising of the eyes. So I try, perhaps rather lamely, to explain that every so often you read a paper in which someone has done something so smart or revealed some hitherto unknown, extraordinary feature of how cells or animals work that you can't help running around showing it to everyone. And yes, occasionally, even in your own lab, you do look down a microscope or get some kind of result and mutter, "Gosh, that is just stunning."

So, while there is no denying that cancer can be one of the most distressing things in life, the challenge and incentive to study cancer has conferred enlightenment and wonderment on a quite astonishing scale.

To end this chapter on one other note of cheer, we might make the rather obvious point that if one in three people die from cancer, two-thirds of us do not. This raises the question of whether the two-thirds are just lucky—

in other words, is the accumulation of cancer-driving mutations a randomly distributed lottery? Alternatively, although it would really be a variation on the being-lucky-in-the-genetic-roulette-of-life theme, do they have a genetic make-up that actually gives them enhanced protection against cancer? You probably know of individuals or families that suggest there may be something in the protection idea. For example, there are people who have smoked all their lives but don't get lung cancer. There are also families like my maternal ancestors. In tracing that side of my family tree I have not found anyone who died either from cancer or from a heart attack. That's pretty unusual and suggests that maybe they had lucky genes, at least in those areas. A consequence of being spared those afflictions was that most of my mother's family lived to what is referred to as "a ripe old age" but, in the way that life has of evening things out, many of them appear to have died painful and lingering deaths as diseases like rheumatoid arthritis undermined their capacity to function. That's rather anecdotal evidence for protection against cancers, but we'll come back to this point when we look at some variants in our DNA that do indeed improve our chances of being in the two thirds rather than the one third.

We've seen that numbers tell a story. For cancer they tell us that different types predominate in different parts of the world and that some can be effectively treated while in the face of others we remain largely powerless. Cancers are mainly diseases of old age: we're living longer and so the cancer burden is set to rocket. We could, however, do a bit more to help. Tobacco use, including "second-hand" and "third-hand" smoking, is the single biggest cause of cancers, and its worldwide effects are increasing. Radiation in the form of sunlight is a major cause of some types of skin cancer that are increasingly common. We are all exposed to radioactivity from the earth about which we can do little except for minimizing the levels of the gas radon in our homes, which causes lung cancer and kills one million people a year. There's no solid evidence that magnetic fields from electrical equipment or from mobile phones cause cancers. From the limited evidence that's been collected, you can drink as much tea or coffee as you like as neither causes cancer and coffee may even protect—but do it in a reasonably hygienic setting because bugs and viruses are a major cause of some forms of cancer, in addition to all the other miseries they inflict. Despite the huge number of studies, there is very little clear evidence linking specific things that we eat with cancer, and the best advice is "stick to a balanced diet"—and don't get stressed! Being overweight or obese increases the risk of many different types of cancer as well as of heart disease, which is a concern as it is an expanding problem. Making sense of either surveys or clinical trials of drugs can be tough so give thanks to The Cochrane Collaboration for providing clear summaries of such things in plain English. While there is a general trend in the developed world to earlier detection and increased survival from many cancers, for reasons that are not entirely clear, the UK tends to be some way from the top of the league.

PART II

GENES AND CELLS

FROM NORMAL TO CANCER

3

A BIT OF DOGMA

ATOMS, MOLECULES, GENES, AND PROTEINS

YOU CAN'T GET THE HANG OF CANCER WITHOUT KNOWING SOMEthing about the molecules that make us what we are and that sometimes go off the rails to end up promoting the disease. So this chapter is about biological molecules; a sub-theme is that it's all pretty easy to grasp. There are two reasons for saying it's straightforward: the first is that nowadays it seems that kids learn about DNA and such things in elementary school, perhaps at the expense of more basic knowledge. These days almost anyone you stop in the street can tell you that DNA is the "genetic blueprint of life." The second reason for being cheerful is that it *is* pretty easy. That's not for one second to imply that we completely understand cancer. We don't, but the barrier to understanding is not the basic processes that make cells work, it's the immense complexity of an animal that consists of a vast number of interacting cells and signals—a point that we'll come back to later. Everything does indeed stem from DNA although it makes up only a quarter of one percent (0.25 percent) of the mass of a cell. DNA is crucial because it encrypts the code of instructions from which cells make proteins by gluing small building blocks together into huge chains. Despite the importance of DNA, it's proteins that do the work, enable us to function, and make us what we are. Proteins make up 18 percent of us: most of the rest (70 percent) is water.

As long ago as 1944 Max Perutz called proteins "the machines of life" because they are required to carry out every single process that occurs in living organisms. Human DNA contains the instructions to build over 21,000 different kinds of these molecular machines that carry out this vast range of activities. They make us what we are: they are responsible for the similarities between us, but they are also the reason every human being is unique. No story illustrates the aptness of Max Perutz's phrase better than cancer, although in that context one might add " . . . and of death." The smallest of changes in the DNA instructions—a mutation—can produce an abnormal protein with the potential to generate a cancerous cell.

We'll begin then with how DNA "makes" protein, what makes proteins "the machines of life," and what a "mutation" means to them. These questions are at the heart of what makes the limitless variety of life on Earth and we'll see that, stunning though the upshot may be, the basic way in which the bits and pieces work is easy to grasp. By starting with DNA, we are indeed beginning at the beginning but, like everyone else, DNA needs a home and we've already mentioned the idea of "cells," the basic elements from which living things are built. We'll do a tour of a cell shortly, and that's important because, in the end, it's cells that misbehave in cancer, even though they have the excuse that they're only doing what their DNA is telling them to do. For the moment, just bear in mind that cells are sacs that provide specialized compartments for biological processes to take place. In animal cells the instruction manual (DNA) lives in an inner sanctum—a smaller sac called the nucleus.

DNA MAKES RNA MAKES PROTEIN

Let's start by running through the way in which the information encoded in DNA is transformed into proteins via an intermediate called RNA. Because this sequence is the molecular basis of all life, it has sometimes been referred to as the "central dogma."

Atoms and Molecules

Molecules are groups of atoms held together by covalent bonds. "Covalent bonds" means, simply, a way that chemicals connect or adhere to each other by sharing electrons and, through this sharing of electrons, maintain a stable balance of attractive and repulsive forces between the individual atoms.

Atoms are the smallest structure that has a defined chemical identity—in other words, that is an element, as opposed to a sub-atomic particle (electrons, protons, and neutrons and their constituents). One way of looking at it is to think of atoms as words and molecules as sentences, an analogy drawn by Robert Schoenfeld in his wonderful little book *The Chemist's English*—and, if you don't think a book mixing English with chemistry could be funny and informative, read it!

A molecule of water is made of two hydrogen atoms joined to one oxygen atom by two covalent bonds (or, as most all of us know: H_2O). Covalent bonds are about one hundred times stronger than any of the other forces that hold atoms together. For that reason, in living systems bond-making and bond-breaking almost always need a bit of assistance. That assistance comes in the form of enzymes: enzymes are (almost always) proteins that bind to molecules, bending their shape and reducing the energy needed for the bonding or breaking reaction so it works more efficiently—a process called catalysis.

DNA

The molecular starting point for all known living organisms is deoxyribonucleic acid (DNA). DNA is a nucleic acid (that is, it's found in the nucleus). In human cells it is split up into shorter segments known as chromosomes—46 of them: 22 are pairs in which one half of the pair has been inherited from our mother, the other half from our father. The other pair of chromosomes determines our sex (X and Y in men, two Xs in women), Y being inherited from our father.

The original discovery of nucleic acids was made in 1871 by the Swiss scientist Johann Friedrich Miescher, working at the University of Tübingen in Germany. He had extracted material from the nucleus of pus cells—white blood cells—but it was Albrecht Kossel at the Universities of Berlin and Marburg who eventually showed that this material contained the four nitrogen-containing bases cytosine (C), thymine (T), adenine (A), and guanine (G), thereby winning the 1910 Nobel Prize in Physiology or Medicine (Miescher having expired before Nobel Prizes were invented). Nevertheless, it was to be more than thirty years before the Canadian Oswald Avery actually showed that DNA was the stuff of life, so to speak.

DNA is a polymer, meaning that many individual units are joined together to form a long chain, just as happens in plastics. In DNA the units are called nucleotides and are joined together by covalent bonds. Each nucleotide has three parts: a sugar, a phosphate group, and one of the four bases (A, C, G, or T). For our purposes the details don't really matter, though they are explained more fully in the glossary for those who want to delve more deeply into the specific wonders of DNA. What is important is that the backbone of a DNA chain is a simple repeat of phosphate-sugar, phosphate-sugar . . . with, critically, the attached bases making a huge string of A, C, G, and T units.

DNA molecules have the well-known shape of a double helix. This structure was worked out by James Watson and Francis Crick at Cambridge in 1953 on the basis of data generated by Rosalind Franklin and Maurice Wilkins at King's College London. The bases poke into the middle of the double helix (a little bit like the treads on a spiral staircase) and weakly pair to bases on the opposite chain. The important bit to understand here is that bases do not pair up randomly: A only pairs with T, and C only pairs with G. That means that if you know the order of the bases in one chain (the sequence), you can work out what it is in the other (e.g., if TAGC is the sequence in one chain, ATCG will be the complementary sequence in the other). The weak pairing of bases occurs by hydrogen bonding, a form of electrical interaction between two molecules that involves a charged hydrogen atom. Individual hydrogen bonds are much weaker than covalent bonds but, collectively, such interactions along the DNA chains make the double-stranded form very stable. Stable but not unbreakable—just warm them up and they'll fall apart. Crick likened the two chains to embracing lovers, separable because the forces making the individuals are stronger than the bonds between them. However, as Watson and Crick famously pointed out in their 1953 paper, if the strands of DNA are separated by breaking the hydrogen bonds, the individual strands can be used as a template from which to copy a new strand, which is the basis of DNA replication when cells divide.

One way of picturing DNA is as beads on a string, each bead being one base in the sequence. There are rather a lot of them—over 3,000 million—the number of bases in one cell's DNA. If you wrote out your genetic code in the same font as this book (one letter per base) it would run to 120,000 pages. And jolly boring it would be, too, bearing in mind that you normally go from chromosome 1 through 22 before you get to the sexy bits (that's X and Y).

The sequence of bases in DNA represents the genetic instructions for making organisms what they are, and it is indeed often described as a "blueprint" or "recipe," neither of which are particularly good analogies. Molecular biology has borrowed "template" from engineering but, as that originally meant a mold, it too is really jargon. Given that the term "genetic code" is nowadays known by almost everyone, "code" is probably as good as anything. In other words, the linear sequence of bases within DNA is read as a code by the machinery of the cell and, in particular, it is used to direct the assembly

The double helix. *Two molecules of DNA form a double helix in which the bases pair with each other: A with T, C with G. The two helices intertwine to give a major and a minor groove that enables other molecules to interact with the bases.*

of proteins. But, as the 1950s drew to a close, a major problem in biology was that, as DNA is enclosed in the nucleus but proteins are made only outside the nucleus, there had to be an intermediate that transferred information between the two.

RNA

In 1961 Sydney Brenner, Matthew Meselson, and François Jacob proposed that the intermediates were "structural messengers" whose "message" was specified by DNA sequences. In a key experiment, Brenner and Meselson showed that this messenger was ribonucleic acid (RNA), first discovered by Elliot Volkin and Lazarus Astrachan three years after Watson and Crick determined the

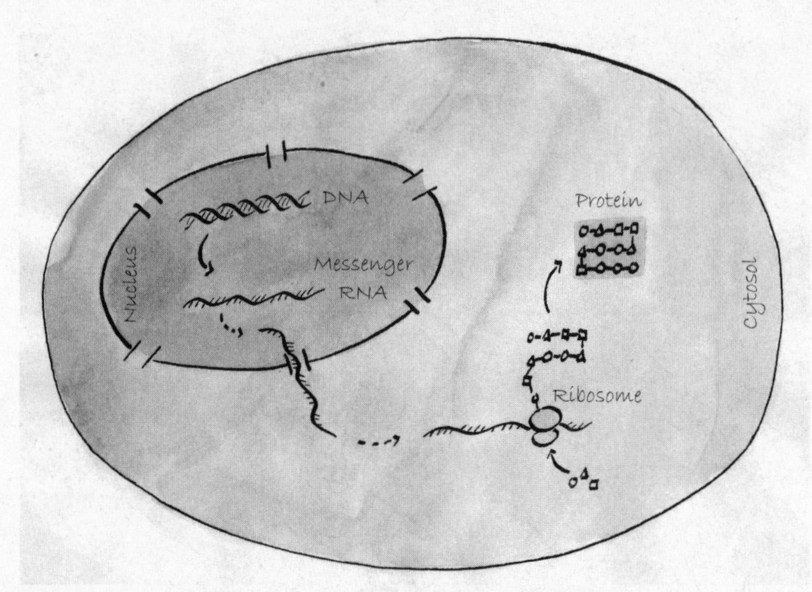

DNA makes RNA (in the nucleus) makes protein. *The DNA code is transformed into RNA. RNA leaves the nucleus via pores in the membrane. To make proteins, groups of three bases in messenger RNA (mRNA) are recognized by molecules that carry the appropriate amino acid for that triplet (codon). The amino acid is joined to the growing protein (polypeptide) chain by a large complex of proteins and RNA molecules called the ribosome (see chapter 5 for a more detailed look at cells).*

structure of DNA. RNA is, of course, also a nucleic acid—a polymer of nucleotides—so it's almost the same as DNA. Almost, but not quite. RNA differs from DNA in three ways: it is single-stranded (it doesn't take up the characteristic double-stranded form of DNA), the sugar has an extra oxygen atom, and one of the four bases is different (thymine (T) is replaced by uracil).

The critical point about RNA, however, is not that it is slightly different from DNA but that its sequence of bases precisely reflects that of the DNA from which it was read. The transposition of the DNA code into RNA is called transcription, and the molecular machine (a group of proteins) that does it is called RNA polymerase. RNA is then exported from the nucleus. Outside the nucleus, RNAs that encode proteins (messenger RNAs or mRNAs) undergo the process of translation.

PROTEIN

Proteins, like nucleic acids, are built up by joining smaller building blocks together, the individual units being amino acids. Translation reads the sequence of bases in mRNA in clusters of three called codons (3 bases = 1 codon). The process is carried out by a sort of "mega-machine," a complex of many proteins and also some RNA molecules called a ribosome. Ribosomes proceed along a molecule of mRNA, "reading" one codon at a time and linking the encoded amino acids together to form a chain. A critical part of each step involves another form of RNA—short molecules called transfer RNAs (tRNAs). These have a three-base sequence (called an anti-codon) that binds (forms base-pairs) with their corresponding codon in mRNA. Enzymes attach the appropriate amino acid to another part of tRNA so that the correct amino acid for the anti-codon/codon interaction is brought to the ribosome to be joined to the growing protein. So tRNAs (there are different ones for each amino acid) are adapters, as originally proposed by Francis Crick.

Proteins then, like nucleic acids, are polymers (sometimes called polypeptides because the links between the amino acid units are peptide bonds, a covalent bond formed in this case when the carboxyl, or proton donor group, of one molecule reacts with the amino group of another, causing the release of a molecule of water). If you're thinking that four bases doesn't sound like much for making all of life, we might inject a mathematical note to the effect that four different bases read in groups of three can generate $4 \times 4 \times 4 = 64$ different triplet "codes." As we make all our proteins from essentially twenty amino acids, four bases is just fine. The fact that sixty-four is bigger than twenty leads to what is called degeneracy in the genetic code—that is, some amino acids have more than one triplet codon. In addition, the code also has room to include some "punctuation" in the form of stop signals that mark the end of a stretch of sequence encoding a specific protein. The critical thing about the chains of amino acids that make proteins is that they are flexible. Adjacent amino acids can rotate about the rigid bond that joins them and this allows the chain to fold up into the three-dimensional shape (conformation) of the complete protein. The linear (primary) sequence of amino acids determines this 3D shape and hence what the protein does.

We've seen that there are more than enough bases, when used in a three-letter code, to tell the ribosome which of the twenty available amino acids should come next. Each of these amino acids has slightly different chemical properties that tell the protein how to fold up into its final shape, and it is this shape that determines what the protein does. The really important question is: How many different kinds of protein can be made? The answer is clearly "a lot" because it is the variety of proteins that makes all the forms of animal and plant life in the living world different from one another. Even so, it's not easy to grasp the staggering variety that can be generated by a system that uses just twenty units to make up all of its individual protein machines. It's not easy to grasp because the answer is, to all intents and purposes, that the variety is infinite. A simple example and a few other numbers will make the point.

Consider just about the smallest protein you could make: three amino acids long. For this you would need three codons in messenger RNA (i.e., nine bases). Each of the codons may specify any of the twenty amino acids, which can be in any order and can include repeats of the same amino acid (so you could have the sequence A-A-A or the amino acids could all be different, A-B-C). Thus it is possible to make $20 \times 20 \times 20 = 8,000$ *different* mini-proteins each containing only three amino acids. In fact the smallest "mini-proteins" (biochemists tend to call very small proteins "peptides") that occur naturally have nine amino acids—three times as long as our three-amino acid example. If you're thinking of having a baby, a bit of the nonapeptide (nine amino acids) oxytocin will come in handy because it will cause your uterus to contract during labor and it will do the same thing to the muscle fibers in your milk ducts so you can do a bit of bonding with junior as well as keeping him off the bottle, at least until he's a teenager. You'll be just fine because your oxytocin gene encodes exactly the right sequence of amino acids out of the 512,000 million possible for a nonapeptide (and yes, that is 512 with nine zeros after it). But nonapeptides are actually tiny by the standards of most proteins. Insulin was the first protein to be sequenced (by Fred Sanger) and at 51 amino acids it is so small that it too is often referred to as a "peptide" hormone. Nevertheless, from 51 amino acids, 20^{51} (that's a bit more than 2 with 66 zeros after it) different proteins could be made! An average-size protein contains about 1,000 amino acids, and the biggest, titin, a muscle protein, has 34,350 amino acids!

There's no point in working out how many different 1,000 amino acid proteins could be made as the answer is beyond comprehension. So it's best just to reflect that from a seemingly simple process, that is joining together just twenty different types of amino acid, an infinite number of different protein "machines" can be made.

How Many Bases?

You might have worked out that if a typical protein has 1,000 amino acids, it would need a 3,000-base stretch of DNA to encode it. And so the obvious question is: Do we have enough DNA to make all our proteins? Well, of course, the answer is "yes" or we wouldn't be here to ask the question—but how much do we have? We mentioned that there are over 3,000 million bases in the complete sequence of human DNA (the human genome), divided unequally between the chromosomes. Altogether that DNA encodes about 21,000 genes, but the coding sequences need only about 1.5 percent of the total number of bases. So our complete genome has plenty of spare capacity when it comes to encoding proteins! A rather pointless exercise, unless you go in for pub quizzes, is to work out how far *all* your DNA would stretch if you put it together in a single string. You have 2 meters per cell and 10^{13} cells in your body—so the answer is that the DNA in your body would stretch to the sun and back *67 times*.

But here's a problem you may not have thought about. How do you get that two meters of DNA into each cell? To make things worse, it's in the nucleus, only three millionths of a meter in diameter, so there's some serious packing to be done. What's more, it's not just a matter of stuffing it in any old how and sitting on the lid. As we've just seen, regions of DNA (genes) have to be "read" to make proteins and, of course, every time cells divide, *all* the DNA has to be replicated. Evolution has solved this problem by adding proteins called histones to chromosomal DNA. To understand how this leads to "more being less," consider DNA as a length of cotton. If you just scrunch the cotton up into a ball you get a tangled mess. But if you use cotton reels (aka histones), you can reduce great length to smaller, more organized blocks. Just to help us grasp what a feat this is, let's scale up DNA to the thickness of a piece of cotton and the length to what you'd need if the nucleus were a golf ball. That's 20 kilometers of DNA cotton squeezed into your ball. Perfect fit, because you've got about 10 million histone "reels" helping, which is just as well because they're all that stands between life and a tangled mess.

3D Proteins

Think of a protein as a string of colored beads—twenty different colors (amino acids), but the total number and their order different for each type of protein. The string folds into a 3D shape that is unique for each protein and that enables it to do its specific work. A lot of them are fairly spherical although, as proteins do everything, they come in all shapes and sizes—cables, bridges, etc. To imagine the effect of mutations, it might help to switch our analogy to a house of cards: the delicate structure represents all the amino acids folded into their working shape. Disturb it and the whole thing may collapse into a useless mess: blowing on the house is like using detergent for washing up—it destroys the 3D structure by dissolving the protein. But, replace one card (amino acid) with a slightly differently shaped piece a bit bigger or smaller than the original (careful!) and the shape may distort without falling to bits. The change in shape may be tiny but, together with the changed chemical activity of the substituted bit, the effect on what the protein does may be enormous.

The idea then seems fairly simple: flexible protein chains are directed by their sequence of amino acids to fold themselves into their working shape. But how do we know about their shape, given that even the biggest proteins are too small to be seen individually? The answer is because of X-ray crystallography, a way of finding out how atoms are arranged in three dimensions within a crystal from the way in which a beam of X-rays is scattered when it hits the crystal. It was from such "diffraction patterns" that Rosalind Franklin and Maurice Wilkins had provided the information that Watson and Crick used to build their DNA model. Another person who played a critical role in that story and who could stake a mighty strong claim to being the most influential biological scientist of the twentieth century was Max Perutz.

Perutz was born in Vienna and it was while studying chemistry at the University of Vienna that he began to realize that the kind of chemistry he wanted to do went on in living organisms. At that time the leading figure in the emerging field of biological chemistry was Frederick Gowland Hopkins, who had founded the subject at Cambridge in 1898 and became the first Sir William Dunn Professor of Biochemistry in 1914. A Nobel Prize winner in 1929 for the discovery of the essential role of vitamins in our food, it was Hopkins who attracted Perutz to Cambridge, but it was at the Cavendish Laboratory—the Department of Physics—that he was taken on as a PhD student in 1936 by John Desmond Bernal. Bernal was in the process of establishing X-ray crystallography as a method for finding out the structure of biological molecules, and it was in his laboratory that Perutz began to focus on the protein hemoglobin to which his career became devoted. That led in 1947 to his becoming head of a Medical Research Council Unit within the Cavendish that evolved into the now world-famous MRC Laboratory of Molecular Biology (LMB), opened in 1962.

Perutz's extraordinary devotion to hemoglobin was eventually rewarded with the resolution of its structure through the use of X-ray crystallography, together with the 1962 Nobel Prize in Chemistry that he shared with John Kendrew. Kendrew had produced the first protein structure of all, for the muscle protein myoglobin. Myoglobin is closely related to hemoglobin—they both carry oxygen and they're both what are called globular proteins, hemoglobin being almost spherical. Hemoglobin is found in red blood cells, and carries oxygen from the lungs. Myoglobin occurs only in muscle (it's what makes meat red), and it binds oxygen even more tightly than does hemoglobin. This means that when blood passes through muscle, lots of oxygen is transferred from the blood to the muscle. Myoglobin has only one chain of amino acids. Hemoglobin has four separate chains packed together, two identical alpha chains and two identical beta chains. The sequences are not particularly close to that of myoglobin. Remarkably, however, the shape of myoglobin is strikingly similar to that of the four units that make hemoglobin, reflecting the fact that both proteins are oxygen carriers.

As the fine detail of the structure was revealed, so too was the way in which hemoglobin works. Hemoglobin can carry four molecules of oxygen, one on each unit. When the first binds, it changes the shape of the unit it attaches to. That movement is transmitted to the other parts of the protein so that the fourth oxygen binds 300 times more easily than the first. This exquisite ballet, driven by oxygen binding, reflects the fantastic flexibility of proteins. The important result is that hemoglobin picks up oxygen very efficiently in the lungs but off-loads it equally efficiently in other tissues where the oxygen level is much lower. It's the communication between the four bits of hemoglobin that makes it such a superb oxygen transporter: it carries almost twice as much oxygen as it would if they worked separately.

The reason for spending a few moments on the molecule to which Max Perutz devoted his life is that a familiar example enables us to perceive just how astonishing proteins are. We've pictured proteins as a necklace of beads folded up to resemble a ball of string. But imagine them too as being alive—breathing—for that is what hemoglobin does when it picks up oxygen. Remember too that all proteins are flexible, adaptable structures. If they were merely molecular statues, they wouldn't work for they would be unable to interact with other molecules to drive biological processes. The exquisite specificity of proteins comes from their amino acid sequence so we should not be surprised that even one error—the mutation of one single amino acid—might have drastic effects, as happens in hemoglobin to cause the relatively common disease sickle-cell anemia.

Mutations

We've already talked quite a lot about mutations, but we didn't define them beyond saying they were changes in DNA. Formally, a mutation is any change in the base sequence in DNA. They come in all sizes from a single change of one base to another (a point mutation) through loss or gain of one base, to huge stretches of DNA being lost, duplicated, or shifted from one chromosome to another. These effects can be caused by chemicals, as discussed in chapter 2 (e.g., the carcinogens that can promote cancer) or they can be the result of a mistake made by the machinery of the cell that it is unable to correct. We often think of mutations in the context of cancers, but they are also the cause of nearly 3,000 hereditary diseases. Those conditions almost always result from a single mutation that gives rise to a protein of slightly altered function. An example is the common hereditary disease cystic fibrosis, caused by mutations in a protein found in the lung: the fault causes a build-up of mucus that affects breathing and can become infected. As well as changing the sequence of proteins, mutations can also alter the amount of protein being made. This "expression level" is highly regulated, and any perturbation is a potential cause of disease. We'll come shortly to cancer-associated mutations.

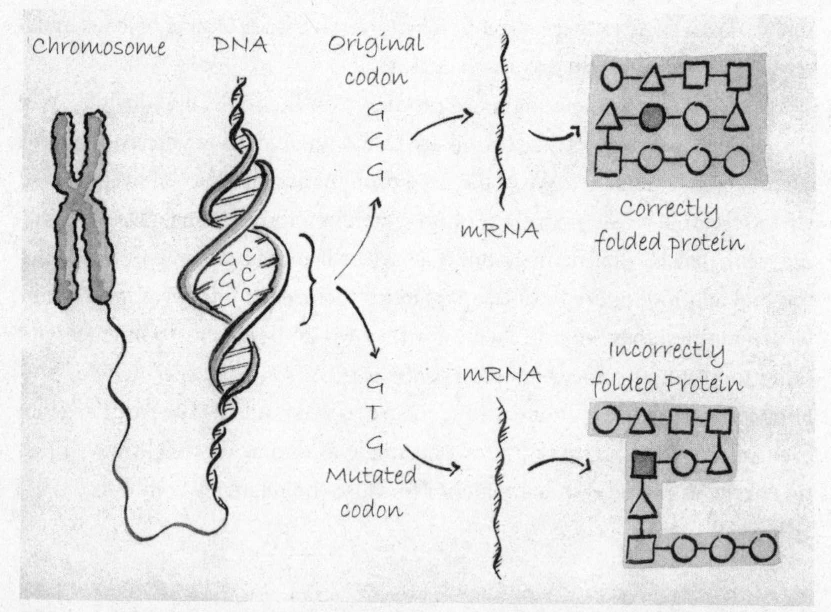

DNA mutations may change the amino acids in a protein. *The effect of a single base change in a codon (a point mutation): the normal codon (GGG) encodes the smallest amino acid, the mutated codon (GTG) encodes a much larger one, the effect of which is to change the shape and action of the protein. This mutation occurs in the RAS gene in about 20 percent of human cancers (chapter 5).*

Keeping It in the Family

As methods improved, it became possible to define protein structures in greater detail than just the path of the polypeptide backbone. It emerged that there are short regions with specific shapes that occur quite often; these are called secondary structures. One such common motif is an alpha-helix (α-helix), which takes a right-handed spiral form. In hemoglobin about 75 percent of the protein has this shape. Other motifs, together with α-helices, contribute to the final shape and hence the function of proteins. Many proteins have an overall globular shape, rather like hemoglobin. Others are almost linear or twist around either themselves or other proteins to form cables: collagen, the most abundant human protein, gives our skin strength, elasticity, and flexibility by forming a triple coil. Proteins can also sit across membranes—essentially two blobs held together by short bits that span the membrane. This suggests that, although there are lots of different proteins (remember that there are 21,000 genes), many fall into groups in terms of structure and, as we've noted, protein shape relates to function. In fact, because the sequence of amino acids determines the final shape, it's much easier to group proteins on the basis of sequence. If proteins have similar sequences, they are said to be homologous, that is, they have similar characteristics because of a shared ancestry. For example, all the members of the globin family (i.e., all the hemoglobins, myoglobins, etc.) have derived from a single ancestral gene over 800 million years. This gives rise to the notion of protein families—groups of proteins that are related by both sequence similarity and function. So protein families expressed in individuals are analogous to species that have diverged during evolution from a common ancestor, a point we'll return to in chapter 5.

It's probably occurred to you that, wonderful though it is to have all these different proteins, controlling what they do is as important as being able to make them in the first place. Just as packaging DNA imposes order in the nucleus, proteins need to be told where to go, whom to talk to, and, if they're enzymes, how hard they should be working. Without proper parental control, there would be cellular chaos. For proteins, once they've been made, by far the most important regulator comes in the form of phosphate, the small chemical group that we've already met as part of DNA. But phosphate can also attach to other molecules and, because it's negatively charged, can have a dramatic effect on the shape, and hence the activity, of a protein. This is nothing to do with size—a phosphate stuck to a protein is like a fly on an elephant—it's the electrical charge it carries that distorts the shape. Where does the phosphate come from? Almost always from ATP (adenosine triphosphate; i.e., it has three phosphate groups), the molecule that's sometimes called nature's energy store because it carries chemical energy around the cell, distributing it to worthy causes (biochemical reactions). Phosphate is shifted from ATP to other molecules by enzymes (called kinases), and that process is called phosphorylation—an event that activates or deactivates many enzymes, telling them to do their job or to stop doing their job. So kinases are enzymes that phosphorylate molecular targets (they're sometimes called phosphotransferases). They're major users of ATP and they are so important that humans have over 500 kinase genes—in other words, about 2 percent of our genes make enzymes of the kinase family. As you would predict in a family that all do the same job (transfer phosphates, albeit to different targets), the members are related through regions of sequence similarity or identity. These "signature" sequence motifs mean that you can place genes in a family even when you don't know anything about what they do.

But you'll have noticed that we've already met ATP as one of the four building blocks of DNA, so ATP is involved in almost everything—a true multitasker molecule. So vital is it that, although we only have a few hundred grams at any one time, we make (and therefore use) our own body weight of ATP every day. The transfer of phosphate from ATP by kinases and its removal by other enzymes (phosphatases) is an immensely versatile mechanism whereby cells can reversibly regulate the activity of proteins. It's the most common form of molecular switch, from which you might also predict that kinases are very important in cancer, and we will indeed meet them in this context in the next chapters.

WHAT IS A GENE? A SHORT WALK ALONG OUR CHROMOSOMES

DNA makes RNA makes protein—a sequence sometimes called the central dogma of molecular biology. In other words, the sequence of bases in DNA, relayed through an intermediate, messenger RNA, tells the machine that makes proteins (the ribosome) precisely what sequence of amino acids should be joined together. The exact code was worked out in the early 1960s by Robert Holley, Har Gobind Khorana, and Marshall Nirenberg, who shared the 1968 Nobel Prize in Physiology or Medicine for their efforts. So this really explained how a gene made a protein. The existence of genes was implicit in

Gregor Mendel's explanation for the results of his pea-breeding experiments, although it took a few more years for Hugo de Vries to come up with a word for the smallest unit of heredity that eventually became "gene," courtesy of the Danish botanist Wilhelm Johannsen—the word from which William Bateson derived "genetics" in 1905.

By the end of the 1960s, therefore, it all seemed pretty straightforward: there were stretches of DNA (genes) that could be turned on and off to make proteins. And, of course, the coding sequences were continuous within any one gene. After all, any different system would be just too complicated to work properly and, anyway, what would be the point? Thus, I suspect, thought almost all scientists at the time and it's therefore hard to overestimate the shock of the finding, published in 1977 by Phillip Sharp and Richard Roberts, that genes are not continuous in their coding sequences but are interrupted by stretches of non-coding sequence (introns). We now know that practically all human genes contain introns. The blocks of coding sequence that they separate are called exons—from "expressed region," meaning that their code is eventually translated into proteins. Therefore, the really critical parts of genes, that is the DNA that encodes proteins, actually come in fragments scattered along chromosomes. When a gene is "switched on," the entire stretch of DNA is copied to make an RNA version of the gene (called the primary transcript). Each intron is then cut out and the exons are stitched together to make a continuous sequence of triplet codons (called messenger RNA, or mRNA).

Control over whether a gene is "switched on"—transcribed into mRNA—is in the hands of a region of DNA that lies before the start of the coding region—called a "promoter." Short bits of sequence in promoters provide sites of attachment for proteins called transcription factors because they determine whether a gene is transcribed or not. You could imagine the DNA sequence of a gene as an airport runway, with promoters as identification lights that mark the area before the touch-down zone. In DNA, however, proteins actually alight on promoters: the combination of these proteins controls the manufacture of RNA from the DNA template.

A gene, therefore, is a region of DNA containing coding sequences (that correspond to the amino acid protein sequence) and non-coding sequences, including the promoter, that control expression of the gene (whether it's made into protein or not). Because we have a pair of each chromosome (apart from X and Y), we normally have two copies of each gene, called alleles (from the Greek *allelos*, meaning "each other").

MICRO RNAS

We've talked quite a bit about "DNA makes RNA makes protein" being a sort of fundamental rule, but a great thing about biology is that if you look hard enough, you'll find some cunning system that breaks any "rule." From that it follows that there are some genes that are transcribed into RNA, but there the rule sticks: the RNA isn't messenger RNA and it doesn't make a protein. We've known about this for a long time because the machinery that actually makes proteins uses some RNA molecules—the protein factory called the ribosome is a huge complex containing lots of proteins but also some RNAs that help the whole thing work.

Quite recently it's emerged that there are a whole lot more non–protein-coding genes. These make RNAs so short you might think they were useless—just 18 to 24 bases in length, which at least makes them easy to christen: they're "micro RNAs"—but they turn out to pack a very solid punch. These short sequences are complementary to sequences in bona fide messenger RNA (that is, they can form pairs with bases in the mRNA sequence in just the same way that DNA forms base-pairs in the double helix). That's unusual for RNA because, as we pointed out earlier, a marked difference from DNA is that RNA doesn't form double strands; the best RNAs can manage is for short regions to curl up on themselves and form loops. However, when micro RNAs bind to their target mRNA, the effect is to stop the message from being processed by the ribosome or, in other words, to jam the protein-making machinery. Sometimes it's just temporarily blocked and sometimes the association leads to the mRNA being broken down, a very effective way of keeping protein from being made from that message.

Given that cells can control gene transcription to regulate whether a protein is made or not, you might be tempted to ask why the cell needs this extra level of control. There are perhaps two ways of looking at this very good question. One is to note that evolution is not an optimization process, it's the emergence and maintenance of things that work by trial and error, and micro RNAs are one of those things.

The other viewpoint is that proteins make organisms: every organism is very complicated and so it isn't surprising that there are multiple layers of control involved in their manufacture, in controlling their activity once made, and in deciding how long they live. One thing is certain: micro RNAs are important because there are over 800 of them in the human genome and they play roles in all the things that matter for a cell: growth, death, and the process of developing from one cell type to another.

It will come as no surprise, therefore, to learn that micro RNAs are certainly important in cancer: the pattern of expression of many of them changes in tumors by comparison with the corresponding normal tissue. Quite often the level of micro RNAs is lower in tumors, which suggests that when switched on, they act as a protection against tumor development. But the patterns of expression vary markedly between different types of tumors. For example, you can distinguish leukemias from solid tumors by their micro RNA profiles and, for some tumors, subsets of micro RNAs may actually be switched on as part of tumor development.

MITOCHONDRIAL DNA

So far we've referred to DNA as though it occurs only in the nucleus of eukaryotic cells—cells that contain within their outer membrane smaller, specialized compartments that are themselves wrapped in a membrane—and that's very nearly true. It would be completely true if virtually all eukaryotic cells didn't have mitochondria, but they do and that's a good thing because mitochondria make almost all of the energy carrier ATP. Mitochondria are like mini-cells within cells: they're enclosed by their own membrane, so they're another specialized compartment with a specific job, as is the nucleus. Their number varies hugely between cell types: some cells have but one (they're presumably snoozing), but the busy chaps (like heart muscle cells that are, hopefully, constantly contracting) have thousands.

Mitochondrial DNA is quite different from the nuclear variety: firstly it's a closed circle of double-stranded DNA, in contrast to being linear in nuclear chromosomes, and there are several copies in each mitochondrion, typically about five. This means there can be up to 10,000 copies per cell. Mitos make their own ribosomes, too, so they are self-sufficient in terms of manufacture. But making ATP is a serious business and it needs a lot more proteins than are encoded by mito DNA. That, of course, tells you that most mitochondrial proteins are encoded in the nucleus—the only other place you find DNA. As we've seen, proteins made from huge strings of amino acids fold up into their compact shapes as they are made in the main part of the cell. In a completely counterintuitive process mito proteins are then unfolded and fed through protein pores that span the mitochondrial membrane, rather like pushing string through a keyhole, before folding up again into their working shape. This making and unmaking costs the cell a small fortune in terms of energy used (i.e., ATP) but, of course, in mitochondria, cells have a bank that never goes bust.

The second difference between mito and nuclear DNA is that mitos use a slightly different genetic code. Not a lot different; just a few codons have a slightly altered function. For example, the AGA codon puts an amino acid into most proteins: in mitos, however, it's a stop signal making the ribosome fall off the message so the protein ends at that point: it's terminated. Not exterminated—just a bit shorter. These two differences have led to the view that mitos were originally bacteria that were engulfed by primitive eukaryotic cells (called endosymbiosis) and the obvious evolutionary gain is of a mighty efficient ATP producer—it's thought that over that time genes have transferred to the nucleus as the mitos have settled into their new home. The less obvious pain is that by-products of ATP generation may help us to get old and to get cancer, and we'll return to that later.

There are a couple more very odd things about mitochondria. One is that they're maternally inherited. That means all your mito DNA came from your mom: the fertilized egg that eventually became you contained about 2,000 copies of mito genome, all but one or two coming from her, and those that came from your dad were rapidly killed. This is why something of a cottage industry has grown up whereby, if you provide a sample, companies will track your maternal lineage from the sequence of mito DNA, in principle back to "Mitochondrial Eve" who would by now be about 150,000 years old. Boys can do a similar thing to get back to "Y chromosomal Adam." Precisely this method has been used to finally resolve one of the great murder mysteries of the twentieth century, the death at the hands of the Bolsheviks of Tsar Nicholas II and his family. In the early 1990s the bodies of all but two of the family were identified by DNA analysis, using samples from living descendants of the Romanovs for comparison. As is well known, the son, Alexei, and one daughter were not identified and, throughout the twentieth century various women popped up more or less plausibly claiming to be Anastasia. In 2007 a second burial site was discovered and sequence analysis of mitochondrial and Y chromosome DNA from those remains finally revealed that all of the royal family had indeed been murdered at Yekaterinburg in 1918.

The other major oddity of mitos is division. How mitos—and for that matter other cell compartments—divide remains mighty mysterious, but mitos do it a lot of the time as the cell they support changes its demands (for example, as a cell grows, it needs more energy). Mitos reproduce by binary fission: they just pinch themselves into two bits, rather as bacteria do. So it's not like the nucleus that divides only when one cell makes two and in doing so ensures a precise 50/50 split of its DNA content. Hence the rather woolly statement just now that there are "about five" genomes per mito. This very odd behavior has some even odder repercussions when you start to think about mutations. You can identify mito mutations that cause an inherited disease because mothers pass it to both daughters and sons, but only the daughters have children with the disease. Because, when they make more of themselves, mitos divide up their genomes in a random way, there is a wide range of severities of a given disease, reflecting the proportion of defective mito genomes in a given tissue.

Over eighty mutations have been identified scattered throughout mitochondrial genes. As you might guess, muscle (e.g., the heart) and nervous tissues that use a lot of ATP are particularly at risk. Needless to say, a number of cancer-associated mutations are known that we'll come to that when we return to the downsides of having a super-ATP factory.

Mitos Make ATP

Hans Adolf Krebs showed how two of the most central biochemical pathways work: the urea cycle and the citric acid cycle (tricarboxylic acid (TCA) cycle or the Krebs cycle). The urea cycle, which Krebs discovered with Kurt Henseleit in 1932, was the first closed loop of biochemical reactions to be identified. It occurs only in the liver and it takes waste ammonia (which is released when proteins are broken down and is toxic—you only need to smell some to guess that it might not be very good for you), and converts it into urea, which is then excreted.

The citric acid cycle takes the products of sugar and fat metabolism and completes their breakdown to release energy in a form that is fed into a sequence of protein complexes called the electron transport chain. All this takes place in mitochondria and results in the production of ATP. This is an amazingly efficient chemical machine because the energy extracted from one molecule of glucose makes about thirty molecules of ATP.

DNA VIRUSES

The word "virus" has us thinking of things that gum up our computers, but they are also the smallest of all organic entities—but they make up for that by outnumbering everything else. Some serious counters at the University of Georgia have estimated that there are 10^{31} virus particles in our biosphere. You may be aware that you have about 10^{14} bacteria in your gut (10 for every other cell that is you), so all told, there are a fair few of them about, but viruses outnumber even the bugs—by about ten to one. We now know that viruses come in a huge variety of types and sizes, but they are all made up of a core of genetic material with a protein coat (called the capsid); some also have a membrane surrounding or within the capsid, rather like the outer membrane of a cell. Regardless of size, viruses are cellular parasites, and they survive because they can stick to and enter cells and then hijack their host's molecular machinery to reproduce themselves. Their variety is bewildering but they can be put into two main categories: those with DNA as genetic material and those with RNA. Tobacco mosaic virus (RNA) was the first to be crystallized to permit X-ray analysis (by Bernal and Isadore Fankuchen), allowing Rosalind Franklin to reveal the first virus structure in 1955.

DNA viruses are responsible for quite a few human miseries including the common cold, influenza, herpes, smallpox, polio, rabies, and chicken pox together with its derivative, shingles. Given the amazing diversity of viruses and the fact that their lifestyle involves throwing something of a molecular spanner into the works, it is perhaps surprising that they are not a major cause of human cancer. Overall infectious agents (viruses, bacteria, and parasites) are thought to cause about 15 percent of human cancers with hepatitis viruses being the biggest culprits, as we saw in chapter 2.

DNA viruses associated with human cancers:

Papillomavirus	Cervical cancer
Hepatitis B virus	Liver cancer
Hepatitis C virus	Liver cancer
Epstein-Barr virus	Burkitt's lymphoma (cancer of white blood cells)

In general, viruses and bacteria do not directly cause mutations in the manner of ionizing radiation and other carcinogens but produce equivalent effects by interacting with proteins in cell signal pathways. For example, the DNA human papillomaviruses (HPVs) that cause cervical cancers knock out a protein that plays a central role in protecting against tumor development. The virus does this by tagging the protein with a signal that tells the cell to break it down. So by the most effective means possible, the virus removes a critical protection against cancer. Hepatitis B virus, responsible for the majority of liver cancers that claim one million lives a year, also makes at least one protein that perturbs normal cell signals to promote cancer.

Burkitt's lymphoma, identified by the Irish surgeon Denis Burkitt in equatorial Africa where it is the most common childhood cancer, arises in a type of white blood cell essential to our immune system and commonly involves the jaw or other facial bones. Anthony Epstein, Yvonne Barr, and Bert Achong at the Middlesex Hospital analyzed samples collected by Burkitt and showed that the disease was caused by a member of the herpes virus family. This became known as the Epstein-Barr virus (EBV: aka human herpesvirus 4). The EBV genome encodes a number of proteins that can overcome the normal growth restrictions on cells, which is thought to reflect the way in which it promotes the human lymphoma. It is also the cause of glandular fever.

CAN RNA MAKE DNA? A RETRO INTERLUDE

Generalizations are extremely useful in science, and in biology in particular. However, they have one drawback: if you look hard enough, you'll find something, somewhere that bucks the trend, as we've already seen with micro RNAs. Thus all organisms stick to the central dogma: whether you're a man or an insect or a plant, your genome is DNA and from that come RNA and protein. Unless you're a retrovirus. We've encountered the viruses that are particularly pathogenic to us and seen that they all have a DNA genome. In fact, being a virus, they don't have much else—they're essentially DNA in an envelope and, because they depend on infecting a host cell to survive (i.e., make more of themselves), they are true parasites. That's why when you are within range of someone who emits an explosive sneeze as a result of a viral infection but appears never to have been introduced to the concept of handkerchiefs, provided you can avoid immediately inhaling a fair number of the million or so viral particles they have fired into the air, you'll be OK because the un-inhaled viruses will quickly die. The second group of viruses, however, are the retroviruses, so called because they have a genome made of RNA.

RNA viruses associated with human cancers:

Human immunodeficiency virus (HIV: AIDS virus)	Kaposi's sarcoma
Human T cell leukemia virus (HTLV–1)	Adult T-cell leukemia

The human immunodeficiency virus (HIV) is a member of the HTLV family (human T-cell lymphotropic viruses). These are RNA viruses and it's well known that HIV destroys cells of the immune system, which means that victims become susceptible to infection. When infection is by the Kaposi's sarcoma–associated herpes virus, the otherwise rare Kaposi's sarcoma (which is not really a sarcoma because it starts in the lymphatic system) slowly develops. Apart from this, however, although many retroviruses can cause tumors in animals, particularly the feline and bovine leukemia viruses, they are rarely associated with cancers in humans.

Like their DNA-containing counterparts, retroviruses have to infect to survive, but they have an additional problem: they can't just hijack the host cell machinery that makes RNA from DNA and protein from RNA. RNA viruses have first to convert their RNA to DNA, which then inserts itself into the DNA of the host cell. After that, it essentially follows the DNA virus method: host machinery transcribes DNA to make RNA for new viruses. So the critical difference is that RNA viruses have found a way to reverse the central dogma. They have done this by evolving the necessary enzyme: a protein that reads the RNA sequence of the viral genome and "reverse transcribes" it into DNA. For a while after its discovery in 1970, this enzyme was called Temin's enzyme after Howard Temin, who discovered it jointly with David Baltimore, working at MIT and the University of Wisconsin, respectively. The enzyme is now universally known by the more helpful name "reverse transcriptase." It was an immense discovery because the viral enzyme works just as well on our mRNA. The ability to convert mRNA back into DNA (a much more stable molecule) provides a snapshot of which proteins are being made in a cell at any one time, and reverse transcriptase has become an essential part of the molecular biology toolbox.

We now have a clear picture of how the DNA code, packaged in genes, defines the proteins that can be made and, through the exquisite precision of the 3D shape of proteins, how it is that a change in the code can have drastic effects. We've also noted that bits of genes can be shuffled in terms of the protein they make, that some genes don't encode a protein and yet control the output of others that do. And we've seen that the cell's power stations have their own DNA, as do some viruses that can have particularly unpleasant effects if they manage to infect us. Next we'll look at how protein machines make cells work.

4

WHAT IS A CELL?

THE SENSATION CAUSED BY ROBERT HOOKE'S *MICROGRAPHIA* REflected the world's wonder at the first glimpse of the miniature world within which all living things are built. It seems likely that its influence reached rapidly across the North Sea to the small town of Delft in the province of South Holland and into the imagination of the proprietor of a successful drapery business by the name of Antonie van Leeuwenhoek. Although an expert on fabrics, Leeuwenhoek had become fascinated by microscopes and, by becoming highly skilled in handling glass, he produced lenses of such quality that he was able to make the first compound microscope (i.e., one using two lenses) and to magnify objects by several hundred times. These advances enabled him to observe animal cells and in 1677 he became the first man to see red blood cells and sperm cells. He also discovered bacteria, becoming therefore the first microbiologist and is indeed called the "Father of Microbiology." This was some fifty years after the physician William Harvey had described the concept of blood circulation in his *De Motu Cordis* (otherwise known as *On the Motion of the Heart and Blood*). As it was then the only conduit for such work, Leeuwenhoek sent an account of these discoveries to the Royal Society, where it was received with great skepticism. It is next to impossible to imagine how people thought before the great revolutions of science because afterwards it all seems so simple and obvious. Clearly the great minds of London had serious trouble grasping the notion of an organism made of just one small cell or that animals and plants might be made of units too small to see with the naked eye.

While thinking Leeuwenhoek to be either a fraud or potty, the Royal Society responded in a way that established a model for how science should be conducted: they sent a team of knowledgeable folk, presided over by a nonscientist—considered to be impartial—to determine, in effect, whether Leeuwenhoek was a charlatan. They duly found that he was a skilled and rigorous investigator who had accurately reported observations that were absolutely reproducible. Thus was Leeuwenhoek vindicated and indeed almost immediately elected a Fellow of the Royal Society (1680). Holland is scarcely renowned for its great scientists, and it's ironic that it is Leeuwenhoek's lifelong friend, the artist Vermeer, who is today remembered worldwide, though few perhaps could name one of his pictures. Even so, it is Leeuwenhoek who changed the world through our perception of things great and small.

Now, of course, everyone knows that all animals and plants are clumps of cells. The number of cells that makes a human clump is estimated to be between 10 trillion and 100 trillion—that's a lower figure of 10 million million and, put another way, the range is 10^{13} (that's 10,000,000,000,000 cells) to 10^{14} cells. Not being sure to within a factor of ten how many cells we're actually carrying around may sound like an admission of scientific failure, but looked at another way it's only one zero in 13 or 14. A friend of mine who ekes out a living as a general practitioner spends his spare time backing horses: he's very analytical and very successful (as a punter, that is) but I doubt that such margins come into his calculations!

One definition of a cell is that it is the smallest subdivision of the body that can live on its own. That is, if you take a piece of tissue, break it up into single cells and put them in a dish with the goodies they need to survive, you can get them to grow and even reproduce themselves. Our cells comprise a sealed sac—the plasma membrane—that contains a gel called the cytosol. This is essentially a chemical soup within which float several more sealed sacs of which the nucleus and mitochondria are the most prominent. Plasma membranes, and indeed all the other membranes in cells, are mainly two layers of fat-like molecules that are long and flexible. This makes membranes extremely flexible, which, in turn, enables cells to undergo prodigious contortions, so that they can squeeze through small gaps or down narrow tubes.

The fatty molecules in membranes are fatty acids: chains of carbon atoms with hydrogens attached (so they're "hydrocarbons"). Pairs of fatty acids link to a kind of molecular bridge, glycerol, that has a phosphate group joined to its third carbon atom, so the whole thing becomes a phospholipid (a lipid is something that's insoluble in water but readily dissolves in organic solvents such as chloroform or benzene). Two layers sit tail-to-tail in membranes that can therefore be called "phospholipid bilayers."

That's all you need to know about the fatty bits of membranes—unless of course in this food-conscious age you're fretting about whether the omega–3 fatty acids in last week's shopping are quietly giving you a heart attack (relax; they aren't).

TALKING TO CELLS

As those of us who do the washing up know, fats are jolly insoluble in water, and this gives rise to a critical property of membranes, namely that they are designed to be more or less impermeable to everything apart from other fatty substances and gases. So that's why membranes are such a good boundary for a cell: they're a very strong, flexible barrier. But for cells to work, things have to cross that barrier and, what's more, there must be movement in both directions—food needs to get in, unwanted stuff must be released. To permit this kind of traffic in a controlled manner, cell membranes also contain proteins that sit astride the barrier, acting as selective channels for molecules or ions—a biological border crossing. These border crossings are essential for basic cell function, but the membrane proteins we're really interested in do a different job. They transmit *information* rather than physical things across the membrane. The signals they deliver determine critical actions by the cell, most importantly whether to divide or not and, of course, perversion of that decision is a crucial permissive event for cancer progression. Picture then a flexible, fatty, double-layered bag speckled with different proteins, most of which span the membrane, providing both nourishment and news, and you can see why the cellular boundary is also sometimes referred to as a fluid mosaic membrane. As well as the carriers and newsbringers, many proteins that do other jobs stick to the surfaces of membranes because that's where their specific activity is required.

The evolution of proteins that poke out of a cell membrane was an essential step along the path from single-celled organisms to animals made up of many cells stuck together. Of course, the capacity to glue cells together implies a change, or perhaps an expansion, in the function of these cell surface proteins from the job they did for the single cell (perhaps only food detection). However, for there to be any development beyond small lumps of cells, the surface proteins have had to evolve much further and develop new activities so that cells with different capabilities could emerge. The upshot of this evolution is the wondrous variety that constitutes the natural world we live in. In humans, as in all living systems, the organs, tissues, and cells that make us what we are, as well as the features that distinguish us both from each other and from other species, are due to the pattern of proteins made. These proteins determine the size, the shape, the location, the color, and the activities that make up the job description for each cell. The astonishing range of cell types that proteins generate may be illustrated by comparing neurons (nerve cells) and blood cells. Red blood cells, and indeed most white cells, are roughly spherical and they're about 5 millionths of a meter in diameter, as you can see from Plate 8. Neurons, on the other hand, are transmitters, sending information by electrical and chemical signaling between their tentacle-like ends. To carry out these roles, nerve cells can be over one meter long.

Most of us are aware that we originate from the fusion of two cells, egg and sperm, and it scarcely requires pointing out that to get to the 10 million million plus cells of an adult, that single fused cell has to go forth and do a good deal of multiplying. This process of cell multiplication, variously called proliferation, replication, or division, is the essence of cancer. The phenomenal precision with which organisms are built implies that cell division normally occurs under the most rigorous controls; it is subversion of these controls by one means or another that characterizes a cancer cell.

A critical feature of a multicellular organism is that individual cells respond to their environment. The decisions that cells have to take—whether to proliferate or to change into a slightly different sort of cell or even to commit suicide—are decided by the group of genes switched on at any time. This pattern of gene expression is controlled by signals that arrive at the nucleus and that reflect the environment of the cell at any moment.

The majority of these environmental messengers come in the form of proteins that are released from other cells in the body and move through the circulation to find their targets, the information transmitters poking out from the surface of cells. One of the best known is insulin, made by cells in the pancreas and released when those cells sense the so-called well-fed state—that is, raised blood levels of the sugar glucose. Certain tissues, in particular the liver, have surface proteins to which insulin in the circulation sticks and it is this interaction of messenger (hormone, which is another name for messenger protein) with specific target (cell surface receptor) that initiates intracellular signaling. Liver cells respond to the insulin signal with a large increase in one type of channel protein on their surface: the channels, of course, carry glucose into the cell where it tops up an energy store.

Shaking Hands with the Messenger

Intuitively, one might suppose that the simplest (and perhaps most accurate) way of getting a signal from the surface of a cell to the nucleus would be for the signal molecule to cross the plasma membrane and then move through the cytosol, enter the nucleus, and do its stuff to alter gene expression. Well, that may be simple, but for the vast majority of hormone signals it's also completely wrong. Hormones (messenger proteins) do indeed dock with specific receptors, but that interaction then switches on a signal sequence within the cell that *does not* require the hormone to enter the cell at all. It's true that hormones may eventually be taken up by their target cells (together with their receptor), usually as a prelude to their being broken down, which is one way of switching off the signal when it's done its job, but that process is essentially independent of signaling.

So how does the news that a hormone has arrived at the cell surface get across the barrier of the membrane? The answer lies in the large three-dimensional structure of the proteins involved. As we've seen, the information receivers span the membrane: they're essentially two blobs of protein at either end of a stick. When the hormone binds, it makes the shape of the receptor change to accommodate it, just as oxygen does to hemoglobin (chapter 3). It's rather like shaking hands with someone: the shape of your hand changes as you grip theirs. The clever bit is that a relatively small change in the blob on the outside of the cell is transmitted to the blob on the inside via the trans-membrane bridge. This signal transmission is helped for most hormones by their being able to bind to two separate (but identical) receptors simultaneously, so it pulls them into an embrace. The fact that the outer parts are drawn together means that the cytosolic domains have to follow suit—in the same way that it's difficult to kiss while keeping your bottom halves far apart. This drawing together of the internal parts of receptors (or blobs, as we've called them) in turn causes them to change shape—not a lot but just enough to act as a signal.

The singular fact therefore about the delivery of most signals to cells is that the messenger does not need to enter the cell: the news of his arrival is transmitted across the membrane by a shape change in the receptor protein. That is testimony to the incredible plasticity of proteins, but the receptors that are particularly involved in transmitting proliferation signals have one more amazing trick.

Spreading the News

We've seen that signals inside cells are switched on when internal "blobs" of activated receptors touch each other—but how? These internal blobs are, of course, proteins, but more than that, they are enzymes. When the receptor is inactive, their shape prevents the enzyme from working, but when the messenger arrives and the receptors embrace, the enzyme is turned on. You'll guess that the enzyme is a kinase (it transfers phosphates), and the groovy thing about these receptor kinases is that they do it to themselves. That is, message arrival results in the receptors themselves becoming decorated with a scattering of phosphates (adding phosphate = phosphorylation), and it is these that serve as "launch pads" for signal relays within the cell. Phosphates are added to the receptor at specific positions where the amino acid tyrosine occurs in the protein chain. These "phosphate receivers" thus facilitate signal transduction. A large, closely related family of transducers work in this way, individual members responding to different messengers, and they're known as receptor tyrosine kinases (RTKs).

These phosphate groups make very sticky binding sites for proteins in the cell that carry the message forward. These proteins fall into two classes: they are either "adapter" proteins or enzymes. Adapter proteins act like bits of molecular Velcro: because they have two sticky bits, they can draw proteins to specific locations or pull two proteins together so that they can talk to each other. Enzymes are recruited to signaling pathways in exactly the same way as adapter proteins—by having the right sticky bit. But enzymes are catalysts, which means that, once activated, they keep on carrying out their specific reaction until they are somehow switched off: this feature means that when there's an enzyme in a pathway, the signal is amplified. Amplification is just what you want to increase the sensitivity of the response and, on the principle that if one is a good thing, two or three is better, signal pathways often have a sequence (a cascade) of enzyme steps that happen one after another. The upshot of such pathways is the movement into the nucleus of phosphorylated proteins that activate specific programs of gene expression.

The system that nature has come up with is rather like the use of runners carrying news from distant parts of a battlefield to central command. In the cell, rather than one runner going the whole way, individuals do a bit of the job, decide they've had enough of being in the firing line, pass on their message to the next chap, and then get their head down. Of course, relaying information by word of mouth is so notoriously unreliable that party games have been devised based on its shortcomings—and it has almost certainly contributed to a good many military debacles. Fortunately, the exquisite precision of proteins means that their conversations are a much more reliable way of passing on information than is usually achieved by the bodies they inhabit.

For this idea of flexible molecules we should be grateful to J.B.S. Haldane, whom we met in the preface. We noted that he was rather a bright spark and, in the 1930s when it was only just being shown that enzymes were proteins,

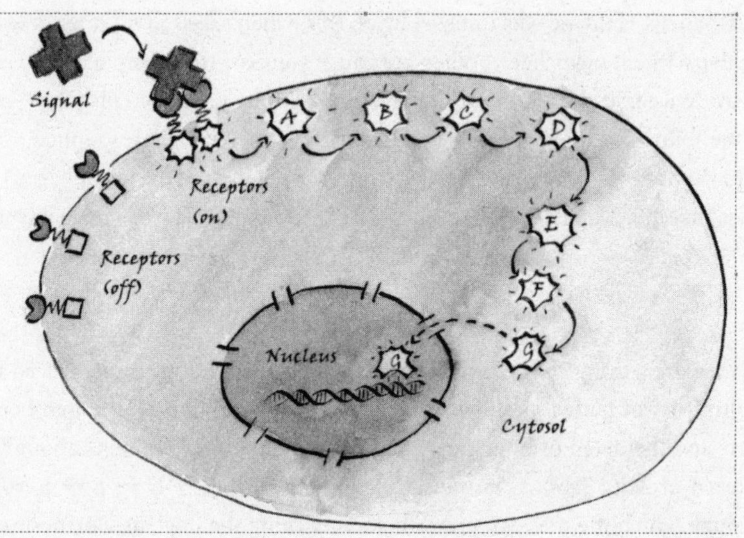

Signaling to cells. *The hormone signal binds to its receptor and activates a protein relay (A → B → C →, etc.) that signals to the nucleus. The receptor is shown as two blobs at either end of a wiggly line—the bit that crosses the membrane. The blobs are a bit like the globular proteins we met in chapter 3. Phosphate groups attached to the internal part of an activated receptor act as launch pads for cell pathways like A → B → C →, etc.*

he suggested that they stuck to their target by interactions that were weak in chemical terms. Haldane argued that, if the contact was through several weak links, the enzyme might be able to distort the shape of its target, making it more reactive (think back to the handshake again). That reactive shape is called the transition state, and his idea of flexibility in molecules is central to how we think today not just about enzymes but about any interactions involving proteins.

Anyone who has worked at Cambridge University should be grateful to JBS for at least one other thing. In 1925 he had an affair with a lady whom he later married. The publicity this generated attracted the attention of a body called the *Sex Viri* ("six men"—six senior members of the university, presided over by the chancellor) who were the guardians of university morals (you couldn't make this up) with the result that Haldane was dismissed for "gross or habitual immorality." JBS, needless to say, contested the case, describing the guardians as "sex weary" en route, and was duly reinstated, thereby establishing that thenceforth the university could not interfere in the private life of its staff. Whew! Haldane was unfazed by all this: when asked in an interview for the department newssheet, "When are you at your worst?" he replied, "There is no evidence that the depths of my potential iniquity have been plumbed" and to the follow-up question "Do you think life worth living?" responded, "Yes, but I do not think the majority of resting bugs, dons, and bacteriophages [viruses that infect bugs] are alive. My answer only applies to higher organisms."

How Many Messengers in a Pathway?

As with everything else in biology, there are no absolute rules and indeed the multiplicity of pathways shows considerable variation in both the number of parts and the detail of how they work. A typical pathway contains about half a dozen proteins (shown in the sketch as A → B → C → D → E → F → G) through which the message passes before reaching the nucleus. Furthermore, some proteins cut out the middlemen by interacting directly with receptors, becoming phosphorylated, and then detaching from the receptor and migrating straight to the nucleus.

How Many Pathways? How Many (Different) Receptors?

Considering, as we have, a single pathway starting from one type of activated receptor, cell signaling looks pretty simple. Nevertheless, there are a couple of complications that we should have in the back of our minds when we think about what makes cells do what they do. The first is that every animal cell has lots of different types of receptor on its surface. Some will be present in very large numbers (e.g., about half a million copies of some growth factor receptors can be present on one cell at any one time), and others will be much sparser. The number of receptors of a given type may vary greatly depending on the state of the cell (how old it is, whether it is dividing, etc.) and the overall complement of receptors depends on what sort of cell you're looking at—in fact, the pattern of specific receptors that a cell carries defines what type of cell it is (e.g., a liver cell or a fat cell). This gives rise to the complication that, in principle, any or all of these receptors may get turned on and thus switch on their own signal pathways to the nucleus at the same time. A second complication is that some of the pathway proteins not only pass on their signal in a linear fashion but can also communicate with proteins in other pathways. For obvious reasons, this is sometimes called "cross-talk" and we'll come back to it later in the context of signaling in cancer cells.

Why So Many Receptors?

Cells, just like the people they make up, have an unquenchable interest in food. A good many of the types of receptors turned on by hormones essentially signal a "well-fed" state and, while such signals may not drive division on their own, they may be necessary for cells to respond to other signals and actually start proliferating. Other receptors may respond to messengers reflecting development or a response to a stress of some sort. A rather obvious example would be the requirement to replace damaged skin cells when you cut yourself.

The overall picture then is that individual signal pathways are easy to understand and that cell signaling is only complicated because of the huge amounts of information the cell somehow manages to process into a decisive response: "I'm going to divide/turn into a different type of cell/commit suicide," etc. Despite our detailed knowledge of the components of signaling pathways, we really have little idea about how cells do this critical decision-making bit.

Imagine a hall full of people, all talking in small groups. Listen to any one group and you understand the information being exchanged but stand to one side and all you hear is a chaotic hubbub. Our cells, however, have the capacity to distil the hubbub of multiple information inputs into one succinct message.

RTKs are probably the most important receptors in cancer because they control signal pathways regulating the growth, location, and lifespan of cells—and it is these pathways that are particularly prone to mutation in cancers. A few other types of receptor signal in different ways and some of these can also contribute to tumor development. All the same, the essential features of cell signaling and how it can go wrong in cancers are illustrated by RTKs and we will stick to them.

One final complication, mentioned only because it will crop up when we talk about how cancers develop: there are some signals that have a negative effect on cell growth. That is, cell growth and proliferation is, like so many other biological systems, a balancing act. At any one time there may indeed be signals telling cells to go forth and multiply, but what happens is the net result of those signals and others that tell the cell to pipe down, not divide, and to remain quiescent or even to commit suicide. The negative signal systems work in broadly the same way as the positive ones: messengers bind to receptors and turn on signal pathways within the cell. There are some minor operational differences, but the important effect is that these signals arrest cell growth rather than driving it.

Too Much of a Good Thing?

Cell signaling should really come with a health warning: if you Google "signaling pathways," you'll shortly find yourself regarding something that looks like a Tokyo subway map; there's an awful lot of signaling that goes on in cells. To keep things simple we've just stuck to one main signaling pathway because once you've got the hang of one, you have a good working understanding of the process, and the details of other types are not really necessary. It's worth bearing in mind that, in some ways, cell signaling resembles the electrical wiring in an old house that has been extended and fiddled with by many owners. New lights are put in, but old cables are left in the walls: phone lines are disconnected but the wires are left—it's too much of a hassle to remove them and the new bits work fine side-by-side with the old. Animals are built in the same way, so there's lots of old junk in their genomes—but sometimes it's not fully disconnected. The spasm that sets off hiccups is something we are susceptible to because the layout of our nerves that control breathing is a residue of our origins as a fish 400 million years ago—a layout utterly useless to us since we climbed out of the briny and got the hang of breathing without gills. Just a mildly irritating side effect of DIY evolution unless you were the chap in the *Guinness Book of World Records* who hiccupped for sixty-eight years when "irritation" probably didn't do it justice. Though thinking about it, perhaps not completely useless: about one in three people who have cancer of the esophagus have had frequent hiccups and difficulty swallowing. This is presumably because as a tumor grows into the trachea, it may throw the hiccup switch by mechanical pressure setting off the involuntary spasm. But don't panic the next time you get hiccups—we all get them—it's only if they persist (last for over forty-eight hours) that there may be an underlying problem.

STEROID HORMONES

Having seen that most signaling molecules, that is, hormones, are proteins and don't need to get into cells to deliver their message, we should mention a second group that work in a completely different way. These are the steroid hormones familiar to all because they include the sex hormones testosterone and estrogen. All steroids are made starting from cholesterol, also familiar because, we are often told, too much of it is a bad thing. Cholesterol is found mainly in plasma membranes: there's roughly one cholesterol for every phospholipid, giving a degree of rigidity to the flexible plasma membrane—which affects how easily things can cross them.

The various modifications that give rise to the steroid hormones convert cholesterol from an essential structural component of cells to a messenger. The major steroid hormones are vitamin D, cortisol (or hydrocortisone), estrogens, progesterone, and testosterone. They're all reluctant to dissolve in water, which means that to be carried around the circulation, they need to attach to something that *is* soluble, the most common carrier being the protein albumin (which makes up about 60 percent of the protein in our plasma, the clear fluid component of blood—blood is only red because of the red cells in the fluid). When these complexes come into contact with cells, the steroid hormone can leave its carrier and diffuse into the lipids of the plasma membrane. By this means, the hormone's signal is delivered to the cell and, in contrast to the protein hormones that require the signaling relay (A, B, C, D, etc.), the messenger actually enters the cell and proceeds all the way to the nucleus to direct gene expression. It can do this because mammalian cells contain specific proteins to which steroid hormones bind.

These are a family of receptors that act in the nucleus to control transcription. Steroid hormone receptors are therefore transcription factors, and they all contain regions that bind to specific DNA sequences in the regulatory regions of their target genes. Activation is via another shape change caused when the hormone attaches to its receptor. Some steroid hormone receptors pick up their hormone in the cytosol and then move to the nucleus; others remain bound to DNA waiting for the hormone to arrive.

Steroid hormones are particularly important in cancer because estrogens can stimulate the growth of breast and endometrial tumors and testosterone may accelerate prostate tumors. For example, estrogen receptors are over-expressed in 70 percent of breast cancers (referred to as ER-positive). Both ER and progesterone receptor status are used to predict response to treatments intended to control the effect of hormones (endocrine therapy). One clear distinction is that ER^+/PR^+ tumors respond much better than ER^+/PR^- tumors.

ONE INTO TWO: THE CELL CYCLE

All of the business with messengers and receptors and intracellular signaling proteins arises from the need to tell the nucleus what's going on outside the cell. Perhaps the most critical information of all is the instruction that the cell should set about growing as a prelude to dividing into two daughter cells. That

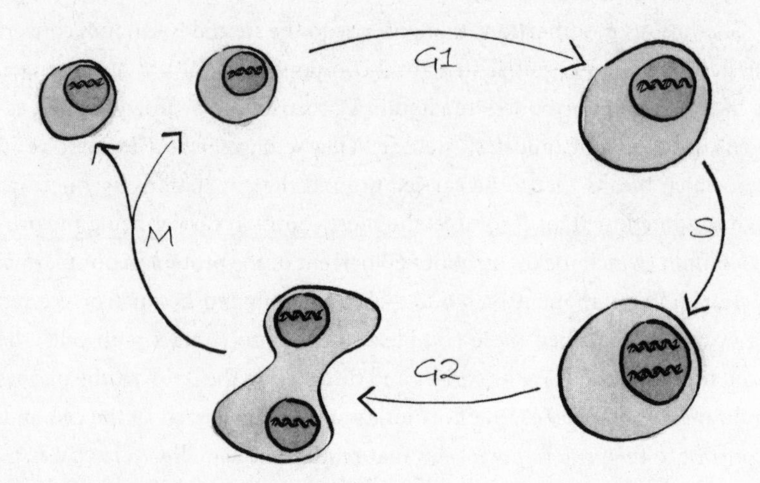

The cell cycle. *DNA is duplicated in S phase, one cell becomes two in M phase, and these events are separated by two growth (G) phases. The main driving force for the cycle is the sequential action of a number of enzymes called kinases that phosphorylate key targets. Kinase activity requires partner proteins called cyclins. Cyclins are made when required and then broken down: this ensures that the cell proceeds stepwise through the cycle.*

may be the instruction but why does it need to go to the nucleus? We know that proteins are the things that do the work and make cells what they are, so why can't they just be told to get on with it? The reason is that for the single cell, just as for the entire organism, the decision to reproduce itself is a major step. It needs to grow a lot—in fact to double its size—and to do that it needs to take up nutrients and generally become much more metabolically active than when it was a resting (quiescent) cell. Most critically, however, it needs to replicate its genome. Each daughter cell will receive identical copies of the DNA sequence that characterizes the organism; doing that requires synthesis of numerous proteins that make up the replication machinery. There are lots of players then but, in principle, the steps involved in cell division are very straightforward: there are two growth phases (G1 and G2, G for "gap") that separate S phase (when DNA synthesis takes place) from mitosis (M phase) when one cell becomes two. The whole sequence is usually called the cell cycle.

Each of these steps requires a specific set of proteins to do the jobs in each phase and to ensure that the cell progresses sequentially from G1 through S and G2 to M phase. That means that the right genes have to be activated to make the RNAs that encode the proteins. To do this, the signaling pathways that control the whole process converge on the nucleus because their end products are transcription factors that turn on the relevant genes. We have already seen how important the process of phosphorylation is in controlling signaling pathways, so it will perhaps be no surprise to find that the most important of these genes encode kinases (enzymes that transfer phosphate to other proteins). In fact, progress around the cycle is driven by the sequential actions of a series of kinases.

This closely related group of enzymes is amazing because it ensures that the cell cycle goes only forward and does not cut corners. For example, it's probably obvious that it would be disastrous if the cell had two shots at replicating its DNA before it divided. If that did happen and the cell subsequently managed to divide, the daughters would have abnormal amounts of DNA and would almost certainly not survive. By phosphorylating appropriate target proteins, the cell cycle kinases provide the driving force for the cell to make the successive transitions between G1, S phase, G2, and M phase. This sequential activation of kinases has one other very clever feature: at the critical steps not only does a given kinase promote the next step but it can target proteins involved in earlier stages of the cycle and block their function. In other words, when one set of proteins has done its job, it can be switched off for the rest of the cycle.

The central role of phosphorylation in the cell cycle emphasizes how powerful this modification is in regulating the action of proteins. It's so powerful in fact that many kinases are themselves phosphorylated by yet other kinases as a way of controlling their own activities. We have already met this type of control in the activation of receptor tyrosine kinases. The kinases that drive the cell cycle require phosphorylation of specific amino acids in their own sequence before they become active. This complex interaction of kinases in controlling signaling and the cell cycle goes at least some way to explaining the 500 plus kinase genes that we have. Because it has to be possible to switch off the cell cycle, it will be obvious that the controls must also include some enzymes that reverse the actions of kinases—that is, knock off phosphate groups. Enzymes that do this are called phosphatases.

Clever though phosphorylation is as a regulatory mechanism, you might guess that, given how all-important cell division is to any organism, there will be additional layers of control. There are. Quite a few in fact and we will come to some of the others later. For the moment let us consider just one, the most dramatic in the cell cycle. In the 1980s a colleague of mine in the biochemistry department at Cambridge, Tim Hunt, had the habit of disappearing for much of the summer. It emerged that he was spending his time at the Woods Hole Marine Biological Laboratory and, as this is situated on the coast of Massachusetts, we naturally assumed that this involved much hobnobbing on the beach with the upper crust of American society, leaving the rest of us to make the best we could of what usually passes for summer in Cambridge. In fact, that coast is also a particularly good place to find clams and sea urchins. These creatures produce large eggs that make very manageable test tubes for biochemical studies. What Tim was really doing was finding a way to see the proteins that were made during the first cell divisions that occur after eggs are fertilized. Fertilized eggs, like most cells, can be grown in the lab provided they are fed. The nutrients must include the amino acids that make new proteins: if one of these is tagged with a radioactive label, any proteins made will be radioactive and, after extracting them from the cells and separating them in a gel, they can be visualized by exposing a piece of film to the radioactivity they emit. It's rather like an X-ray except that the film detects particles emitted from the tagged proteins rather than radiation that's passed through your dislocated shoulder.

One day late in the summer of 1982, Tim appeared in the department gleefully waving a rather battered piece of film and telling anyone who would listen how important it was. What the film showed was a rhythmical appearance and disappearance of radioactive bands that were specific proteins coming and going as newly fertilized cells went through their first division cycles. Tim was right to be excited even if most of the rest of us were slow to see the point: the discovery led to his sharing the 2001 Nobel Prize in Physiology or Medicine with Lee Hartwell and Paul Nurse. Because the levels of these proteins fluctuated, Tim called them "cyclins." It turned out that cyclins attached to the kinases that drive the cell cycle; in addition to being correctly phosphorylated, kinases must be associated with their partner cyclin in order to become active. These kinases therefore are called "cyclin-dependent kinases" (CDKs).

So the two basic control processes that regulate cell division are phosphorylation, both by and of specific CDKs, and a progressive synthesis and breakdown of cyclins that also regulate CDKs. They are so fundamental to living systems that the major genes involved have remained virtually unchanged during evolution—you can swap yeast cell cycle genes for their human equivalents and the cells will still divide quite happily.

Given that some cells are dividing all the time in animals, it may seem counterintuitive for the process to involve something as drastic as destroying key bits of the machinery. Notwithstanding the energy cost in making new proteins, there is clearly no better way of ensuring that something works only when needed than wiping it out the rest of the time. This has emerged from evolution as such a powerful strategy that we encounter it not only in every molecular pathway in cells but also in the behavior of cells themselves. The elimination of cells in a precisely controlled way is a requisite part of normal development, and it's a major protective mechanism against cancer. We'll come back to this later, but the idea is that when cells get stressed, as happens if their DNA is damaged, one way out is for the cell to commit suicide for the good of the organism as a whole.

A LITTLE BIT OF CULTURE

It almost goes without saying that to find out how individual cells work, you need to be able to take them out of the animal to which they belong and grow them in the laboratory, as was done in the discovery of cyclins. That is to say, cell behavior is determined by complex interactions of protein pathways and, to disentangle those pathways (and find all the bits we mentioned in the last chapter), you need to be able to grow a substantial number of cells of the same type so you have something to experiment with. Walk into just about any biology lab in the world and you'll find freezers and big Thermos flasks of liquid nitrogen full of small vials of frozen cells. These are "cell lines" that were originally established by taking a piece of animal or plant tissue, breaking it up into individual cells, usually by mild treatment with enzymes, and then giving the cells the appropriate nutrients to keep them happy (salts, amino acids, and hormones) in a sterile culture dish. Usually this "culture medium" is a liquid that covers cells stuck to the bottom of the dish: sometimes cells just float around in it. Once you can grow cells in culture, not only can you watch how single cells behave but you can break them up and ask what the individual bits (mostly proteins) do. If you're very adventurous, you can try putting bits back together to reconstitute some of what makes a cell (e.g., the machinery that makes proteins). One of the cell lines lurking in freezers around the world has gained a certain degree of prominence, namely HeLa cells, first isolated by George and Margaret Gey in 1951 at the Johns Hopkins Hospital from Henrietta Lacks, a patient with cervical cancer.

With a bit of luck, cells will grow at least until they fill the surface area available when they can be removed, again by gentle treatment with an enzyme (which just chops off the sticky bits by which cells attach to surfaces), and either frozen for storage or diluted into a new culture dish to produce yet more cells. Cells can also be grown on jelly-like semi-solid or solid growth media such as broth or agar.

This method of "tissue culture" or "cell culture" has, in one way or another, provided the basis for almost every advance in biomedical research in the last fifty years. The term "tissue culture" is sometimes also used in the slightly more specific sense of literally growing pieces of an organ in the lab. Originally, this meant removing a section of tissue and growing it without attempting to digest it into its individual cellular components. Very recently this term has been re-focused in the context of the use of stem cells to generate entire organs and also in the development of methods to grow pieces of skin from patients to provide grafts for their burns.

These methods are often described as *in vitro*—literally "in glass"—and indeed sterilized glass was still the only type of vessel available for cell culture when the most dramatic advances were made during the development of the first vaccine for poliomyelitis. This was Jonas Salk's inactivated poliovirus vaccine, first used in 1952. A critical requirement for developing the vaccine was to be able to grow live poliovirus in the laboratory so that it could then be chemically inactivated for use as a vaccine. Three scientists working at the Children's Hospital in Boston, John Franklin Enders, Thomas Huckle Weller, and Frederick Chapman Robbins, had managed to do this in 1948, first in cultured human cells and then in a line derived from African green monkey kidney cells, from which the virus was ultimately made. For this achievement they shared a Nobel Prize, though not with Salk, despite the fact that his vaccine led to the elimination of infantile paralysis: he had to be content with being one of *Time* magazine's 100 Most Important People of the Twentieth Century.

THE SECRET OF IMMORTALITY

Quite apart from its contribution to the polio vaccine, the culture work of Enders, Weller, and Robbins really laid the foundation for modern cell biology. But there is a bit of a problem in growing cells that we haven't yet mentioned but that is extremely relevant to cancer. The problem is that if you follow the earlier instructions—take a bit of animal tissue, break it up into individual cells, and cover them in a medium—what happens is that most cells grow quite happily and divide to make more daughter cells. However, after they've done this a few times, they start to die and in the end, almost all the cells die. Normal human cells grown in culture divide about fifty times, gradually expiring. This was first noted in 1965 by the Philadelphia-born Leonard Hayflick, and the barrier to continuous reproduction is thus called the Hayflick limit. The cells that haven't died but have stopped being able to divide have undergone what's called "replicative senescence"—they've reproduced until they're too old to do it any more—and how long that takes varies depending on the type of cell and the animal it came from. It occurs because at each cell division a bit is lost from the ends of the chromosomes: this region is called the telomere and an enzyme (telomerase) exists that can prevent that loss but telomerase is more or less switched off in normal dividing adult cells. Whatever its value or cause, the Hayflick limit presented a real problem for scientists because their primary cell cultures vanished as they worked with them. Having to start again at regular intervals by taking fresh tissue is not only a nuisance, particularly perhaps to the creature donating the tissue, but has the added disadvantage that you cannot be sure you have exactly the same type of cell. The first immortal cell line, derived from mouse tissue, was described in 1943, and a further breakthrough that permitted the establishment of the numerous cell lines mentioned above came from George Todaro and Howard Greene in 1963. They isolated mouse embryo cells, grew them in culture, and showed that indeed almost all of them stopped growing as they approached their Hayflick limit. Just occasionally, though, some cells would survive this "crisis point" and start growing again at a fairly rapid rate. Moreover, these cells could be unstuck when they completely covered the culture dish, diluted, and allowed to continue growing, a process that could be continued indefinitely. Todaro and Greene had shown that normal cells with a finite life span could give rise to "established cell lines" that survived indefinitely.

A huge number of *in vitro* studies have subsequently been carried out: virtually all drugs or reagents used for human therapy rely on cell culture in their production and testing. Of course, these days there's not much glass involved; it's almost always disposable plastic. Plastics have been a wondrous thing for cell culturers but they add a huge sum to the running costs of labs.

Although much less feted than, say, the discovery of the structure of DNA, these advances were also one of the foundations of modern molecular biology. And, as so often happens in science, they were in turn built on much earlier observations that were in many ways ahead of their times. The first recorded example of culturing bits of tissue was in 1885 by the German zoologist Wilhelm Roux, who kept a piece of chicken embryo alive for several days by immersing it in saline solution. The same Roux, incidentally, had provided support for Theodor Boveri's ideas by suggesting that different chromosomes could mutate independently—in other words, that changes in DNA occurred randomly.

The "saline solution" that Roux used was developed by his contemporary Sidney Ringer, whose remarkable experiments launched, albeit with some delay, even more than the field of tissue and cell culture. Ringer was professor of medicine at University College London and, among other things, showed in frogs that the poisonous effect of aconite (the member of the buttercup family, also called wolfsbane, that makes aconitine) was the result of its paralyzing the ventricle of the heart, an effect that he could reverse by applying atropine. Although it would be many years before they were identified, what Ringer had observed was the effect of the two agents on a class of cell surface protein receptors—the muscarinic acetylcholine receptors. Atropine (commonly extracted from deadly nightshade—*Atropa belladonna*), famously used by Cleopatra to dilate the pupils of her eyes, also attaches to these receptors but instead of switching them on (activating), it stops them from signaling. It does this by blocking aconitine binding and in the eye its effect is that you stay wide-eyed. So next time you are seduced by such a gaze, you may reflect that you are observing a competitive antagonist that is stopping the circular muscle round the iris from contracting. Taken orally, atropine is therefore mighty poisonous but, as Cleopatra's primpings showed, it penetrates the skin quite easily—just touching the leaves of wolfsbane can be dangerous. Despite this, aconites are one of many plants with a long history of use in Asian traditional medicine, albeit usually in somewhat detoxified form.

Labeling Proteins and Coloring Cells

Being able to grow cells in the laboratory was one of the great advances in biology and it has underpinned many of the discoveries about cancer. It has been complemented by the triumph of being able to 'see' proteins in cells and therefore to measure how much is present and to track their movement as cells respond to signals. The first step in this extraordinary achievement happened in 1962 when Osamu Shimomura, Frank Johnson, and Yo Saiga isolated a photoprotein—that is, a protein that can emit light—from luminescent jellyfish. They called it aequorin because the flashing free-swimmers (they're the things that look eerily like spaceships floating in the ocean) belong to a group called *Aequorea.* They obtained aequorin by squeezing the proteins from the luminescent bits of jellyfish (to give "squeezates"), but they also found another protein that gave off a greenish fluorescence and helpfully called it green fluorescent protein (GFP). Fluorescence is a form of luminescence in which light is generated by some sort of chemical change rather than by heat. Fluorescence occurs when a molecule absorbs a photon (the elementary particle that carries electromagnetic radiation) and this absorbed photon causes the molecule to emit another photon of lower energy (lower energy = higher wavelength). To cut a substantial story short, it transpired that when calcium binds to aequorin, the protein glows blue. However, some of this blue light is absorbed by its companion GFP and re-emitted as green light (lower energy). In due course, other creatures were also found to make GFPs (*Obelia,* a sort of jellyfish and *Renilla,* a sea pansy).

From this discovery about what goes on in jellyfish came two of the most dramatic and important advances in the history of cell biology. One of these was the direct measurement of the level of calcium within cells. The other turned GFP from something various marine animals use in self-defense to a flag that could, in principle, tell you where any protein in any cell was at any one time.

Jellyfish belong to a group of over 9,000 species that includes corals and anemones called cnidaria—with a silent *c* and an equally silent capacity to sting the unwary. Which is pretty amazing but the really stunning thing is indeed the capacity of some to emit light. By and large, proteins don't do that. In other words, the amino acids that are strung together to make proteins in us and almost all other organisms are mostly not fluorescent (in fact, three of the twenty amino acids fluoresce very weakly, just enough to be a nuisance when doing cellular fluorescence microscopy but not enough to make us glow in the dark). Not so GFP, which has a specific sequence of three amino acids that takes part in a complex reaction that makes it fluoresce. So that means that there is a gene that encodes a light-generating protein in the genome of luminescent jellyfish.

In chapter 8 we'll trace the story of genomic sequencing and describe what's meant by "cloning" genes; for the moment, suffice it to say that Douglas Prasher tracked down the GFP gene, enabling Martin Chalfie to engineer DNA that could be taken up by an animal of choice, which then made GFP. Chalfie, after acquiring a PhD from Harvard, had worked as a postdoc on worm development with Sydney Brenner and John Sulston at the Cambridge Laboratory of Molecular Biology and then moved to Columbia University, still working on the worm. With this background, the model animal was obvious, and he assembled his GFP-coding DNA with a regulatory sequence that would switch it on only in one type of worm cell. So the world first saw the use of GFP as a marker for gene expression in the form of a "glow worm" with green fluorescent spots in the few neurons where GFP was made. This meant that there was nothing special about jellyfish except that they had been smart enough to evolve the GFP gene: if you put that gene in another animal (including us), cells will make the protein and it will glow green when excited by light of the right wavelength.

Jellyfish are smart, therefore, but that's not to say their efforts can't be improved upon, and in 1995 Roger Tsien made the first mutant of GFP with enhanced fluorescence (brighter light). Tsien was an American who studied chemistry and physics at Harvard before gravitating to the UK to do a PhD at Cambridge, returning to the US in 1982 to work at the University of California. Tsien's "molecular engineering" of GFP led to the generation of a number of other mutants with different spectral properties—giving, for example, blue, cyan, red, or yellow fluorescence.

Once you've cloned the gene, there are two main ways in which GFPs (normal or mutant) can be used: (1) by inserting the GFP gene next to the regulatory region (promoter) of a cellular gene of interest (as Chalfie had done in his worms); this means that when the cellular gene is "switched on" by its promoter, the artificial promoter-GFP gene will also be activated and fluorescent protein will be made separately; and (2) by fusing the GFP gene with the gene encoding the protein you wish to study so that a fusion protein is made (i.e., your protein with GFP tagged on the end). Ideally, this will be made up of the normal protein carrying the GFP flag wherever it goes in the cell. The availability of other fluorescent proteins means that, in principle, you can engineer the simultaneous expression of different proteins within cells, each carrying its own fluorescent color.

For these methods to work, aside from piecing together the bits of DNA that encode the promoter, the protein you're interested in, and GFP—you need to be able to either get them into cells to study effects *in vitro* or express them in live "transgenic" animals. These methods are almost routine nowadays and are used in labs all around the world. If you make a transgenic animal in which GFP is switched on in most cells, you get something that glows under fluorescent light. So you can make green mice and now you can buy fluorescent zebrafish (GloFish) online. For making all this possible, Osamu Shimomura, Martin Chalfie, and Roger Tsien shared the 2008 Nobel Prize in Chemistry.

FURRY FOOTNOTE AND A FISHY BIT

Humans and mice have a common ancestor from which we diverged about 75 million years ago (recall that the earth is 4.5 billion years old (4,500 million), the first simple cells appeared about 3.8 billion years ago and multicellular life started 1,000 million years ago). Since that separation 75 million years ago, mice and men have gone their separate ways in superficially spectacular fashion so it's perhaps surprising that our genomes have remained about the same size (3.2 versus 2.6 billion base pairs) and encode about the same number of genes (20,000). It's a somewhat humbling fact that mice have slightly more genes than we do (about 1,000) although, seemingly, they use most of that number to refine their sense of smell, rather than to enhance their intellect. The mouse has counterparts to about 99 percent of our genes with an average similarity in sequence of 85 percent. DNA sequence identity with our closest living relatives, the chimpanzees, is over 96 percent, so mice are indeed a more distant branch of the family. Nevertheless, over 90 percent of our genes occur in the same order in our genome as that of their mouse counterparts—enormously strong evidence that we do indeed share a common ancestor.

So, although it may not be as immediately obvious as our relationship with apes, especially to those whose reaction is to hop onto the nearest chair at the sight of small furry objects, mice are quite close relatives. This has turned out to be critically important because, aided by the fact that mice breed, well, like rabbits and are easy (although very expensive) to maintain in the laboratory, they have contributed a substantial portion of what we know about cancer biology. In view of one of the big differences between us, namely that the average mouse life span is about two years, that may seem surprising. After all, cancers are diseases that generally develop slowly on the time scale of human life. Nonetheless, as mice have pretty well all the genes we have, they can reproduce most of the genetic abnormalities that drive cancer in us. What's more, it has transpired that, despite their short lives, mice can develop spontaneous cancers either as a result of inbreeding or after the experimental introduction of specific genes into their genome, that is, as transgenic mice. In addition, mice have been used as models for cancer development and treatment by transplanting human tumors after suppression of their immune system so they can't reject the tumors.

With all that going on, it's not surprising that mice have something else in common with us: they get stressed. How do you stress a mouse? Well, aside from the more ghoulish ideas you might think of, mice and indeed rats resemble humans in being jolly gregarious souls so, if you put them in cages on their own (what the psychology fraternity call "social isolation"), they become much more timid and frightened. If you stress them a bit more (by restraining them or wafting in the smell of a predator) they get even more stressed and for longer periods than do their mates living in a group. How do we know? Well, as you might guess, we do the same thing as the Trier Social Stress Test experiment did for humans: measure their cortisol levels—which are indeed higher for longer in the loners. You will also have guessed where this is leading: not only do isolated rodents have disturbed hormones, they are much more prone to cancers. Female rats living on their own, for example, are over three times more likely to develop malignant breast tumors than those living in groups. That's a staggering increase, and it seems likely that it happens because the abnormal stress hormone levels affect the expression of some critical cancer genes that we'll meet later.

Humans and mice aren't the only creatures that get cancer; it's also a major cause of mortality in domestic and farm animals. About 10 percent of cats and dogs develop some form of cancer during their lifetime. Dogs in particular show a similar range of spontaneously occurring cancers to humans. The development and pathology of the tumors is also similar to human cancers and so is the response to treatments adapted from the corresponding human therapies. Dogs can also be the victims of inherited cancers; for example, German shepherd dogs born with a mutation in the folliculin gene develop something called RCND (that's renal cystadenocarcinoma and nodular dermatofibrosis, for dog owners). This occurs because the second (normal) allele of folliculin is lost during postnatal development. Humans can suffer the corresponding condition when they are born with a defective folliculin gene—called Birt-Hogg-Dubé syndrome—that causes benign skin tumors and is associated with an increased risk of kidney cancer. It's such a rare condition though that it is counted as the number of families afflicted—about 100 worldwide. Many animals including dogs and cats can develop tumors formed from mast cells, which are part of the immune system and help repel parasites (they store histamine, which, when inappropriately released, causes allergic reactions). These skin tumors, called mastocytomas, vary in severity from benign to highly malignant.

Nonetheless, some very strange things do happen in dogs that, as far as we know, don't occur in us. Canine transmissible venereal tumor (CTVT: also called Sticker's sarcoma) has spread to dogs on five continents. Analysis of DNA from these tumors shows that they don't develop from cells of the host animal (i.e., they don't arise from scratch in the afflicted animal) but are all derived from a single original tumor cell that developed between 200 and 2,500 years ago. They are, in effect, a cell line that has been propagated for longer than any other known. This tumor was first identified in 1876 by a Russian veterinarian, Mstislav Aleksandrovich Novinsky, and it mainly affects the external genitalia. The most unusual feature of this cancer is that it is spread either by coitus or through licking, biting, or sniffing, so that live tumor cells are picked up by the recipient animal. One or two other examples of contagious tumors have been found, notably a particularly virulent facial tumor in the Tasmanian devil. These carnivorous marsupials are reckoned to have the strongest bite of any living mammal and it's through biting that the cancer is transmitted, which happens quite frequently as the males especially are mighty unfriendly chaps. The tumors can grow quickly and end up preventing the animal from feeding so this cancer, derived like CTVT from one common precursor, is threatening the survival of the species.

Although many tumors can be transferred between animals in the lab, these examples in nature are very rare, and there is no evidence that tumors can be transferred between humans. The only partial exception to this rule happens very rarely when tumor cells are passed from mother to fetus, that is, to an unborn baby still in the womb. The cancers are melanomas or leukemias with unique "driving" mutations that we'll come to in the next chapter. For the moment, the important point is that cancer is not transmitted between humans after they are born so that, for example, you cannot get cervical carcinoma or prostate cancer by cell transfer during sex.

Zebrafish

You may not have expected fish to feature on the cancer menu but the fact that they are coming to the surface just shows how wonderful biology is. The zebrafish (*Danio rerio*) is a kind of Himalayan minnow, noticeable because of five horizontal blue stripes on its sides. Biologists have found them attractive not for their prettiness but because they have turned out to be a very good model for studying embryo development (their embryos are large, grow outside the mother, and are transparent). They're also of great interest because they can regenerate most bits of their body after injury. Zebrafish stripes arise because of the way special cells that give skin and eye color are arranged, the pigment-containing cells being called chromatophores. These enable the fish to adapt their coloration to the light level to which they are exposed because the pigments in different classes of chromatophores determine the color that they reflect. The "chroma-" bit simply means color, "irido-" being iridescent, and "melano-" black or brown. The zebrafish's melanophores are the equivalent of human melanocytes mentioned in chapter 2 (though we don't have any of the other types of pigment cell). Chameleons are perhaps the best-known creatures that have chromatophores because of their ability to redistribute the pigments within their cells so that they change color, for example, when they're frightened or they fancy a spot of mating.

Cancer biologists have recently joined the ranks of zebrafish enthusiasts as a result of the generation of two mutant forms. One (the nacre mutant) is unable to make any melanocytes while the second cannot make iridophores, has few melanocytes and translucent skin (the Roy Orbison zebrafish: I keep telling you scientists are repressed comedians). When these two are cross-bred, double mutants (for both nacre and roy) are produced that have no melanocytes or iridophores and, somewhat bizarrely, are almost totally transparent. These see-through fish, named casper for their ghostly appearance, have been used as a model for melanoma using cells taken from a mouse tumor. The cells are inoculated into the tummy or the blood and five days later black masses of melanoma cells are visible by eye within the body of the fish (Plate 9). As with mice, these fish don't provide everything you need in a model cancer system but they do enable us to see directly tumors growing inside a living animal.

CELLS IN SUMMARY

Cells are the smallest unit of life—meaning that they can reproduce themselves. A human being is a collection of many billions of cells and each responds to messages from its environment so that it can do its job. These messages come in the form of hormones, usually proteins (e.g., insulin), that talk to a cell by sticking to other proteins poking out from its surface. That surface is a membrane that is both the cell's boundary and a selective barrier. The most common way of getting information across that barrier is through spanning proteins that slightly alter their shape in response to messengers. This permits an info-flow to the control room of the cell—the nucleus where DNA resides—and that happens through a relay of proteins. The other main class of messengers, steroid hormones, can diffuse across membranes and thus travel all the way to the nucleus—but the upshot is the same as for protein hormones: they regulate whether genes are "switched on" or not.

The most critical decision for a cell is whether to reproduce itself. To tell it to do so, the incoming signals impinge on the cell cycle, a process in which cells grow, double their DNA content, grow a bit more, and then split in two. The machines that control the cell cycle are proteins, and the most common way of regulating their activity involves subtly modifying their structure by attaching a phosphate group. Other proteins—enzymes—do this, and those that transfer phosphate groups are called kinases. Control of the cell cycle is obviously central to the well-being of animals, and it lies at the heart of cancer. You might predict, therefore, that evolution has come up with several layers of regulation to minimize the risk of losing control. The most dramatic of these is a class of proteins called "cyclins"—because they come and go during the cell cycle—and their sequential appearance and specific interactions with kinases ensure that cell division proceeds in an ordered, stepwise manner.

Most of what we know about cells has come from being able to grow them under artificial culture conditions in the laboratory. In particular, this has led to the discovery of many "cancer genes" and to the understanding of how mutations affect the activity of the encoded proteins. We're now poised to look at some critical mutations in cancer genes before going on to see how they affect signaling in cancer cells.

5

"CANCER GENES"

WHAT ARE THEY?

IF YOU'VE GOT THIS FAR, YOU'RE IN GOOD SHAPE. REALLY. YOU KNOW enough cell and molecular biology to hold your own in the cancer field, which is good going considering that students spend years doing degrees in these things. So, given that you're happy with genes and proteins, what makes a cell and how it interacts with the world, what else do you need to know before we look at how cells change when they become cancerous? Just a bit more detail really: we've seen that to process incoming signals, cells have relays that channel information from receptors to the nucleus—that's where the decision making goes on. Because cells are hit by many signals, they have lots of different types of receptors to pick them up, and these can turn on a waterfall of signal pathways. What's more, some of these pathways talk to each other (cross-talk) inside the cell en route, so to speak, to the nucleus. Why do they do that? Well, they seem a bit like spectators at a sports match: some cross-talk is positive, tweaking up the signal strength in the target pathway (cheering on your team), some is the reverse (essentially telling the opposition to pipe down or the referee that he's less than perfect). So there's a babble of information flow and somehow—science still does not have a completely clear understanding of cellular communication—the cell refines this into a specific response or even multiple simultaneous responses. Because the critical signal pathways control growth and death, you might think that *any* abnormal part could contribute to cancer if it didn't kill the cell first. In principle that's true but, fortunately, it looks as though only a rather small number of genes can mutate into major cancer "drivers."

There's one thing to get clear at the outset. Despite the chapter title, there's no such thing as a "cancer gene." Cancers come about because changes in the DNA code (mutations) affect how proteins work, or even whether they are made at all. The proteins in the cancer front line are the ones that carry out the key jobs of cellular life: they control growth and division, where cells live, and when they die. Perturb any of those and you may have a recipe for cancer. The big problem with science-speak occurs when it escapes and pops up in the popular press or on TV. How often have you heard "the media" refer to "the breast cancer gene X"? To those who are not too scientific, this conveys the idea that you somehow pick up a gene that will give you cancer rather as you might get food poisoning from a visit to a slightly dodgy eatery. Wrong. These genes, which can cause so much trouble when they're damaged, are not only perfectly normal parts of our molecular make-up, they're absolutely essential to life. It's just that something has gone wrong somewhere along the way, and they've turned into a bit of a delinquent.

Where shall we start with the explanation of cellular delinquency? Well, quite often in science it's helpful to retrace key events in chronological order because they really built the foundations of the subject—and doing it with hindsight means you don't go down all the blind alleys that seduced your predecessors. The first cancer genes were discovered in viruses. Starting from those and using just three or four examples, we'll see that almost any contortion of the DNA code that you could imagine happens in cancers to make mutated proteins that uncouple the cell from its normal constraints. These mutations are almost always picked up as we wander through life. The exceptions are the unlucky few who get a bad genetic deal at birth in the form of a mutation that makes them very cancer-prone.

MAKING "CANCER GENES"

Tumors result from subversion of the processes that control the growth, location, and death of cells; we know that this loss of normal control happens because mutations change the way that proteins work. In principle, you could imagine two extreme versions of this idea. Make too much of a protein or change one of its units (amino acids) and you could have a hyperactive effect: a much more efficient signaler. At the other extreme, if a gene becomes damaged or is completely lost, the encoded protein can no longer be made. It turns out that almost all cancers are driven by combinations of these two types of mutation and, to distinguish between them, genes that become abnormally active are called oncogenes, and those that lose function are tumor suppressor genes.

As you'd expect, the proteins made from these genes control the growth and multiplication of cells and things like how sticky they are—which in turn affects how easily they can move around. Because tumor suppressors lose function in cancers, it is obvious that their normal role is to hold things in check. We've met this idea before, and these negative regulators of cell growth are another manifestation of the recurrent theme of biological balancing acts, this time between accelerators and brakes in cell proliferation. To avoid confusion, the normal, unmutated forms of oncogenes are called proto-oncogenes (the prefix "onco-" means "tumor" from the Greek *ónkos,* mass or bulk: "proto-" comes from the Greek *prótos,* first).

The current estimate is that of the 21,000 genes we have, there are about 500 oncogenes and 100 tumor suppressor genes. This figure is slowly creeping up, and a reasonable guess is that we're going to have to deal with about 1,000 "cancer genes."

There are two more groups of "cancer genes" that don't precisely fit into the categories of oncogenes or tumor suppressors. The first are the micro RNAs that we met blocking the production of proteins in chapter 3: anything with that kind of influence might well affect cancers. The other group act as a sort of maintenance gang repairing damaged DNA; we'll return to the importance of this activity at the end of this chapter.

AN EARLY BIRD . . .

It is arguable that the first experiment in cancer molecular biology was carried out right at the beginning of the twentieth century. Its author was a Texas-born physician called Peyton Rous who had studied at Johns Hopkins Medical School in Baltimore before becoming a research scientist at the Rockefeller Institute for Medical Research in New York. Making the acquaintance of a Plymouth Rock hen that had developed a large tumor, Rous cut out the growth, ground it up, and filtered it. He used the finest filter then available, one that would have held back everything down to the size of bacteria, and injected what passed through the filter into normal chickens. To his surprise, these birds developed exactly the same type of tumor as the original. Now Rous, of course, had no idea what he had done in molecular terms, but he was sufficiently shrewd a scientist to comment in his 1911 paper that the agent carrying the tumor might be "a minute parasitic organism."

Rous is one of the very select band of scientists who will always be remembered by name for their deeds. Nevertheless, he had to wait a while for immortality to be conferred and probably holds the world record for the longest time between seminal discovery and award of a Nobel Prize—which came in 1966; he was the guy who shared it with Charles Brenton Huggins (chapter 1).

The inordinate delay in recognition suffered by Rous was because he had to wait for technology to catch up with his experiment. The advance was electron microscopy that can magnify objects up to 2 million times, and it showed that what Rous had transferred to give his chickens cancer was in fact a virus. So that was why it had slipped through his fine filter: viruses are about twenty times smaller than bacteria—and he was right: viruses *are* minute parasitic organisms.

The realization that Rous had generated a tumor by injecting a virus into his chickens started the hunt to discover just what it was that had such drastic effects on its host. By that time, the advances we traced in the last chapter meant that cells could be grown in the laboratory, so it was a relatively simple experiment to infect a "culture" with Rous's virus and watch what happened. The result was startling but perfectly consistent with Rous's findings. We saw that when "normal" cell lines are grown in culture dishes, they fill up the available space and then stop growing. In that state they look not unlike ripples in the sand after the tide has gone out, at least if you view them at low magnification. A few hours after infection with the Rous virus, however, they go through a remarkable transformation. The infected cells unstick themselves from the dish and contract, literally pulling themselves together to begin dividing again. As they do so, their surface membranes ruffle, taking up the slack, and they assume a frilly appearance that is characteristic of a tumor cell growing in culture. The process of turning normal cells into tumor cells is called "transformation," and the question then became "What was carrying the transforming power of the virus?"

Viruses are parasites because they hijack the machinery of the cells they infect to make more viruses. But even biology isn't perfect and sometimes odd bits of the host's DNA get picked up by the viruses. It's a risk the viruses run for using the underhanded tactic of trying to pass off their genetic material as belonging to the host. For Rous's virus the extra bit encoded a complete protein—the cause of the tumors in his chickens all those years ago.

The gene that had been picked up by Rous's virus was initially dubbed "sarc" (now SRC) because the tumors it caused in his chickens were sarcomas, and it eventually emerged that the protein it encoded was a kinase, one of the family we met (in chapter 4) regulating the cell cycle by shuttling phosphate groups between molecules. The real shock for the scientific world, however, was that the "sarc" kinase specifically phosphorylated just one type of amino acid—tyrosine. The surprise was caused by the fact that it was known that most tyrosines were *not* phosphorylated and, suddenly, here was this protein that could generate cancer cells by doing something that no one had previously thought at all significant: adding phosphate to tyrosines. It took some time for the penny to drop that the act of tyrosine phosphorylation was a signal: the molecular equivalent of throwing a switch that turned on a circuit telling the cell to start proliferating. In other words, it short-circuited the normal control of cellular behavior and pushed the cell toward becoming a tumor. An electrical engineer would immediately have recognized the significance of the almost negligible amounts of phosphorylated tyrosine in normal cells: a burst of activity from a tyrosine kinase dramatically increases the level, giving a high signal-to-noise ratio that leaves the cell in no doubt as to what it should do.

THE AGE OF ONCOGENES

The dissection of Rous's virus prompted the hunt for other tumor-promoting viruses, setting the stage for the insight of J. Michael Bishop, who started to work on a virus called MC29. Bishop was a somewhat reluctant graduate of Harvard Medical School who eventually found his true vocation as a cancer molecular biologist and became, arguably, the leading figure in the oncogene field. Bishop knew that MC29 causes carcinomas in birds. Because carcinomas are by far the most common form of human cancers, Bishop reasoned that the genome of MC29 might therefore be the bearer of something at least as exciting as SRC. It was a brilliant piece of educated guesswork because he was able to show that MC29 carries the MYC gene (so named because the virus causes tumors in *my*eloid *c*ells, a type of bone marrow cell) and subsequently MYC has emerged as the most frequently activated oncogene in human cancers. It's difficult to resist quoting Bishop's words: "By means of accidental molecular piracy, . . . viruses may have brought to view the genetic keyboard on which many different causes of cancer can play, a final common pathway to the neoplastic phenotype"—neoplastic being a fancy word for tumor-like, and a phenotype being an observable characteristic. They turned out to be spot on, with the rider that they revealed only a very small fraction of the number of keys that can switch cancer on.

Viral Oncogenes Are Pirated Versions of Normal Genes

Gradually, more and more viruses were found that had acquired cancer-causing genes. In every case the virus was one of the family that have RNA as its genetic material (the retroviruses that we met in chapter 3). With these discoveries there came a problem: Did oncogenic retroviruses have anything to do with human cancer? In other words, it seemed possible that, although viruses with genetic add-ons caused cancer in some animals, that might just be because they were a sort of viral oddball that, though interesting, didn't really tell us anything about human cancer. This concern seemed the more real because retroviruses, by and large, have no effect on humans, as we've noted. Of course, we would not in this story have devoted so much time to the life of retroviruses if it had not transpired that their oncogenes are very relevant indeed to human cancer. The first indication of the truth came in a 1976 paper from Bishop's lab showing that normal mammalian cells have a gene corresponding to the tumor-promoting bit of the Rous virus; they speculated that it might be involved in "the normal regulation of cell growth." Bishop has subsequently described this as "pure bravado" rather than speculation because identifying SRC DNA sequence did not mean it was part of a "proper" gene. Proof that it was took a little longer, but eventually they were able to show that the gene was used to make protein; that was, of course, the normal version of the Rous virus tyrosine kinase.

In due course, all other retroviral oncogenes were also shown to have been acquired from normal genes. As more and more species were examined, these genes turned up in everything from humans to flies to worms. They were, in short, highly conserved across species, a strong indication that they played critical roles in cellular life. In the final piece of this jigsaw, the bravado of Bishop and his colleagues was fully justified: the reason proto-oncogenes, from which oncogenes are derived, are so highly conserved is that the proteins they encode are involved in controlling cell proliferation.

This was a critical moment in cancer biology, and for their discovery of the cellular origin of retroviral oncogenes J. Michael Bishop and Harold E. Varmus shared the 1989 Nobel Prize in Physiology or Medicine.

The First Human Oncogene

The key experiment showing that the genes some viruses have picked up can themselves cause cancer, independent of any virus, came in 1983. In principle, it couldn't have been simpler: extract DNA from a human tumor and chop it into short lengths that get taken up when you add them to cells grown in the lab. The tumor DNA kick-starts some of the cells to divide abnormally and these form clumps that can be picked out. Inject those cells into mice, wait for a tumor to grow, cut that out, and isolate the gene responsible. It's easy to show that the cancer-promoting gene comes from the original tumor DNA and by this kind of experiment Robert Weinberg, Channing Der, and their colleagues identified the first human oncogene, RAS. In fact, the gene had already been identified, as you might guess, in viruses that, when they infect rats, cause tumors—rat sarcomas; hence the name RAS. We now know that there are three closely related human RAS genes, and mutations convert one of them from proto-oncogene to oncogene in about 20 percent of human tumors.

WHAT TURNS A NORMAL GENE INTO A "CANCER GENE"?

We have already paid homage to the extraordinary prescience of Theodor Boveri in reasoning that scrambled chromosomes might upset cell division and hence lead to cancer. Coloring tumor chromosomes reveals that spectacular large-scale shuffling of DNA does indeed occur. Being able to tag a stretch of DNA with a fluorescent label and track its travels in the chaos of the cancer genome would have seemed magical to Boveri, so goodness knows what he would have made of the astonishing revolution by which complete DNA sequences can now be obtained. We'll come to the significance of those shortly, but for the moment the critical point is that, in addition to the shunting of great chunks of DNA, cancer cells contain huge numbers of mutations, typically tens of thousands. What's more, the disruptions feature everything you could possibly imagine happening to a string of letters, all in one cell. If you think about that for a moment, you might conclude that the really amazing thing is that cancers happen at all. The genetic mayhem is rather like your stunning Lego model of the Eiffel Tower being the target of a child's tantrum, reducing it to a jumble of unconnected blocks. From such a mess, how can a cell cobble together enough of the "blueprint for life" to enable it to carry on—still to be able to function? Of course, a lot of DNA changes have no effect, so they don't matter, and neither do those that are fatal for a cell: it will just die, the best form of cancer protection. If that sounds like quite a high price to pay for an insurance premium, recall that we lose about 100,000 million cells a day in the normal course of events, so the odd extra to get rid of a potential cancer cell is a mighty good investment (see chapter 7). Even so, the instability in their genomes means that cancer cells walk a slender tightrope between suicide and survival as cellular assassins.

Let's turn then to the disruptions that occur and how they affect the molecules that make cells—proteins. Having said that just about any kind of derangement you can think of happens, it's helpful for our sanity that, broadly speaking, mutations fall into three types.

The smallest disruption possible is a single base change: this may have no effect at all but, if it happens to make a critical change in a protein or in how much of it is made, there may be dramatic consequences—we mentioned cystic fibrosis in chapter 3. Putting that simple message into code (and remembering that the DNA code comes in groups of three), the triplet GGG inserts the amino acid glycine into the growing protein: mutate a G to T (to give GTG) and a different amino acid goes into the protein. Instead of a switch, a coding unit may be lost altogether or an extra one may be inserted (or deleted) by mistake. Imagine a sequence of triplets . . . GGG AGG TCG . . . An error deletes A, giving . . . GGG GGT CG . . . The "reading frame" is shifted so that the code now makes an abnormal protein from the deletion point onward. Much larger deletions can also occur, resulting in only part of the normal protein being made, which can endow it with dangerous properties, as we shall see.

A quite different type of mutation is a kind of DNA shuffling that, in essence, makes a new gene. That's called chromosome translocation—segments of DNA shift within or between chromosomes to make either (1) a normal protein under different control so that it is made in the wrong place or at the wrong time or in inappropriate amounts; or (2) a completely new protein by fusing coding regions that were originally far apart.

In the most extreme mutations, entire genes or even whole chromosomes may be duplicated or lost. That means that either you end up with lots of copies of a gene—so you get much more protein made when those genes are turned on—or genes disappear completely, so that the proteins they encode can no longer be made at all.

The easiest way to see how those effects work is to look at them in action in three or four major "cancer genes."

Changing bases: Minimal mutations in molecular switches: RAS

1: . . . ATGACGGAATATAAGCTGGTGGTGGTGGGCGCCGGCGGT
GTGGGCAAGAGTGCGCTGACCATCCAGCTGA'TCCA . . .

2: . . . ATGACGGAATATAAGCTGGTGGTGGTGGGCGCCGTCGGT
GTGGGCAAGAGTGCGCTGACCATCCAGCTGA'TCCA . . .

Can you spot any difference between the two rows above? No? Congratulations. Unless you're an escaped molecular biologist, you've just done your first bit of sequence analysis. And you got it wrong. Does that matter? Well, yes if you're one of the 20 percent of those with cancer where the cause is a mutation in a RAS gene. The two lines of letters are the first 75 in the DNA code for RAS so, keeping in mind the triplet code, that's the first 25 amino acids. Count in groups of three until you get to twelve and you'll see that the top line is GGC but the bottom is GTC—that's mutant RAS. How on earth can one letter make a potential death sentence? To answer that we need to recall the delicate structure that is the 3D shape of a protein.

We've already met RAS as the first human tumor protein to be found. It also appeared anonymously in the protein relay (A → B → C →) that talks to the nucleus. RAS normally plays a critical role in this pathway because it works like a switch: it's either "off" or "on." The switch is "on" when RAS binds to a small molecule called G3P—but in fact, though not by name, we've already met G3P as one of the four building blocks of DNA supplying the base G. Here G3P is moonlighting by doing a completely different job: making RAS change shape just enough to open a sticky site for the next member of the signal relay team. It's a splendid example of the way in which evolution has proceeded by trial and error, picking up and adapting whatever comes to hand: unintelligent design, one might say.

Normal and mutant RAS. *A single amino acid change locks this molecular switch in the "on" position: it cannot convert G3P to G2P, and the signal pathway it controls is permanently active.*

What turns the RAS switch off? As well as being able to talk to other proteins and bind G3P, RAS is an enzyme. It chops a phosphate off G3P making G2P. This simple change flips the switch off. That is, when RAS-G3P becomes RAS-G2P, the 3D shape of RAS changes just enough to mask the sticky site for the next protein in the chain. Now you can guess the effect of the one-letter change in mutant RAS: it swaps the smallest amino acid glycine for a bigger one at a critical site—with disastrous results. G3P binds to that site but the change blocks the switch to G2P: RAS remains locked in an embrace with G3P. The molecular switch is permanently "on"—as is a major cancer signal pathway.

You may now wish to retire to a darkened room to meditate upon the slender thread by which human life is suspended.

Missing bits: Deaf to the world: EGFR

1: . . . CGACCCTCCGGGACGGCCGGGGCAGCGCTCCTGGCGCTGCTGGCTGC
GCTCTGCCCGGCGAGTCGGG'CTCTGGAGG . . .

2: . . . ——————————————————————
——————————————————GAGG . . .

OK. Quite enough philosophy for one science book. Now, what do you make of these two codes? Great chunk missing from the second one? So if, say, it was a receptor, maybe it's lost the bit outside the cell that the messenger sticks to, and then perhaps it behaves very oddly. Brilliant! We've made a sequence analyst out of you! These codes *are* from a receptor (we met these shaking hands with messengers in the last chapter: EGFR is the target for a protein called epidermal growth factor (EGF)). And you're absolutely right that having a big piece missing causes some unexpected effects. We pictured receptors as

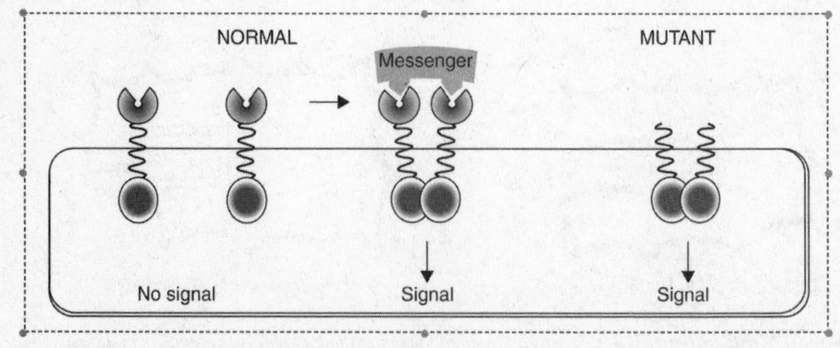

Signaling by normal and mutant receptors. *Mutation activates the signal without a messenger.*

two blobs of candy on either end of a wiggly stick crossing the cell membrane. Messengers pull pairs of receptors together, turning on the parts inside the cell so they start signaling. The mutation in the second line of code indeed takes out the messenger-binding site. This is important for some breast cancer sufferers because that's the mutation that drives their cancer. The odd thing is that losing the outer bit of the receptor turns it on, just as if the messenger was there all the time. That's a great cancer driver when the signal tells cells to grow and divide.

Patching proteins: Chromosome translocations make novel proteins

1:... GCTGTCCTCGTCCTCCAGCTGTTATCTGGAAGAAGCCCTTCAGC
GGCCAGTAGCATC...

2:... CGATGGCGAGGGCGCCTTCCATGGAGACGCAGATGGCTC
GTTCGGAACACCACCTGG...

3:... CGATGGCGAGGGCGCCTTCCATGGAGACGCAGAAGCCCT
TCAGCGGCCAGTAGCATC...

Tricky or what?! Now you have two starter sequences; how do you get your mutant? Maybe you've got there by now: the front half of 2 shifts to the back half of 1 in a chromosome rearrangement—a pretty amazing thing because you'd think it was next to impossible that big chunks of DNA could re-sort themselves yet manage to end up making a sensible message (i.e., a protein). It's a bit like shuffling a pack of cards, laying them down and finding you have a straight flush in spades immediately followed by one in hearts: ♠AKQJT♥109876. Pulling a trick like that would certainly give you a short half-life in Vegas! Freaky it may be but such chromosome rearrangements do occur and they're particularly common in cancers that arise in the bone marrow where red and white cells of the blood cells are made (Plate 8). These are leukemias and lymphomas, different translocations being associated with particular types of leukemia. Put one of these fusion genes in mice or zebra fish and they develop the same disease, showing that the novel protein created by the DNA rearrangement is sufficient to trigger the cancer. The best known chromosomal translocation was identified in 1960 in a case of chronic myeloid leukemia (CML) by Peter Nowell and David Hungerford working in Philadelphia (so it became the Philadelphia chromosome). Some years later Janet Rowley showed that it came about when a piece of chromosome 9 displaced part of chromosome 22, producing a shortened, hybrid chromosome. This DNA shuffle moves a bit of a gene called BCR to a position directly in front of a gene called ABL1 (the sequences above are correct, by the way). This produces a fusion protein (BCR-ABL1)—the driving force for leukemia development. ABL1, like SRC, is a tyrosine kinase enzyme, and the effect of having BCR stuck on its front end is to switch it on permanently. As we've seen, tyrosine kinases can act as powerful proliferation signals, and ABL1 appears to be particularly potent in this respect in bone marrow cells. The Philadelphia chromosome occurs in over 90 percent of CML cases and in 10 percent to 20 percent of patients with acute lymphoblastic leukemias (ALL).

The leukemias have been particularly revealing in terms of cancer development because single cells carrying "driver" translocations, e.g., BCR-ABL, can be identified. Fluorescently labeled probes can detect translocations, so that individual leukemic cells can be selected from a blood sample. DNA from these cells can then be sequenced to determine the complete mutational pattern in individual tumor cells. This approach has been applied to the rare condition of childhood ALL. Three quarters of children with ALL have a translocation that fuses two genes—TEL and AML1. The break occurs in one cell in the fetus that then multiplies. In fact, this type of gene shuffling is not as uncommon as you might suppose, given the relative rarity of ALL. About one child in every one hundred has a TEL-AML1 fusion but only one in one hundred of the children with this fusion gets leukemia—clearly showing that the fusion is a "promoter" event, following which further mutations are required. This is consistent with the disease appearing at widely varying ages, from birth up to about 14 years of age.

Being able to look at the DNA in different leukemic cells taken from the same patient reveals something else that is striking and central to cancer. The patterns of mutation differ between different cells. This tells you that the tumor is made up of distinct groups of cells (i.e., clones), even though they have all started from the same TEL-AML1 fusion event in one cell. These "genetic snapshots" reveal extraordinary diversity in mutational patterns. For example, a clone may acquire an extra copy of a gene and then subsequently lose it as it evolves into a different, albeit closely related, clone. This picture of genomic plasticity can only be visualized by studying single tumor cells, which you can select in leukemias by fishing for the translocated driver gene. The approach has revealed what Mel Greaves, who works at the Institute of Cancer Research and is a leading authority on childhood leukemia, has called the "Clonal Architecture" of tumors whereby individual clones are present in different proportions, each having its own capacity for self-renewal. In other words, tumors start from a single cell with its hand of mutations but they grow into a mutational mixture. This is sometimes called "tumor heterogeneity," which means that treating cancers with drugs is a real pain because you're trying to hit lots of targets rather than one—and the targets are changing all the time.

These conclusions are consistent with several studies of identical twins that originated from a single fertilized egg, where both carry the same initiating mutation but only one has developed ALL. The normal twin has the initiating fusion gene (TEL-AML1) but has none of the copy number changes in the afflicted child. As happens in the game of genetic roulette, the unaffected twin has failed to pick up the extra mutation(s) necessary to complement the "driver" and launch the expansion of tumor clones.

It's not possible to do this kind of single cell analysis in solid tumors, but it would seem almost certain that they too will turn out to consist of multiple clones, modulating their mutational profile independently and proliferating at different rates as a consequence of that profile. In other words, as we might have suspected all along, cancers are a form of dynamic Darwinism.

Chromosome Translocations: Normal Protein, Abnormal Control

A slightly different form of chromosomal rearrangement produces not a novel protein but a normal protein under different control. That is, the shuffling involved places a gene under new management—it becomes regulated by different DNA controllers. This means that, although normal protein is made, it may be in the wrong type of cell, or in the right cell at the wrong time. A quite well-known example occurs in Burkitt's lymphoma (chapter 3) where the gene involved is MYC. Sometimes translocated MYC may also pick up mutations that do affect its function or the half-life of its mRNA, but the main feature of this type of mutation is that the normal gene (MYC in Burkitt's lymphoma) finds itself regulated by a DNA sequence that usually controls a completely different gene.

Multiplying genes

. . . CTGGATTTTTTTCGGGTAGTGGAAAACCAGCAGCCTCCCGC
GACGATGCCCCTCAACGTTAGCTTC . . .
. . . CTGGATTTTTTTCGGGTAGTGGAAAACCAGCAGCCTCCCGC
GACGATGCCCCTCAACGTTAGCTTC . . .
. . . CTGGATTTTTTTCGGGTAGTGGAAAACCAGCAGCCTCCCGC
GACGATGCCCCTCAACGTTAGCTTC . . .
. . . CTGGATTTTTTTCGGGTAGTGGAAAACCAGCAGCCTCCCGC
GACGATGCCCCTCAACGTTAGCTTC . . .
. . . CTGGATTTTTTTCGGGTAGTGGAAAACCAGCAGCCTCCCGC
GACGATGCCCCTCAACGTTAGCTTC . . .

Seems to me that, with all this spending on the National Health Service, doctors ought to help out by teaching genetics to their patients. They could start by replacing those annoying eye test charts they insist on hanging on their walls that have a huge A on the top line then rapidly vanish into meaningless rows of miniscule letters. "Now Mrs. Miggins, for your eye test I want you to read the letters in 8 point Gill Sans Light on the wall at the end of the corridor. It's the first 75 bases of MYC gene DNA. I realize that you may have memorized the normal sequence for this test, so I've selected that from a patient with Burkitt's lymphoma that has a single base mutation—that's what we doctors call a point mutation. Failure to pick this up will mean you'll have to have our newest spectacles. They're $500. Away you go."

The duplication of entire genes occurs frequently in human cancers and was first discovered in one of the most common cancers in children—neuroblastoma. In a search for novel cancer genes, the one that emerged was NMYC. This was a striking finding because, as its name suggests, NMYC is a close relative of MYC, the gene discovered by Bishop that we now know is essential for cell proliferation. Since then it's emerged that most types of cancer overexpress MYC—that is, they make a lot more MYC protein than normal cells—presumably because it is such a powerful proliferation driver. In some colon cancer cells, there are over 100,000 molecules of MYC whereas there are fewer than 1,000 in normal cells. The most obvious mechanism for overexpression is gene amplification, probably arising from mistakes in the DNA replication step of the cell cycle. The number of duplications can range from a mere doubling through the five shown above to the 1,000 or so copies sometimes found in neuroblastomas.

Amplified genes can tell us something about the likely progression of a cancer. In 20 percent of primary breast cancers, one of two receptors on the tumor cell surface are amplified—EGFR or its close relative ERBB2—so that there are multiple copies of each rather than the normal two. ERBB2 amplification is a bad sign because it generally means the tumor will grow quickly (it's "aggressive"). As you might guess from what we've said about MYC, an even worse scenario is high levels of both ERBB2 and MYC, which happens in about 12 percent of breast tumors. ERBB2 is the target of Herceptin, an effective treatment for about 20 percent of early stage breast cancers. We'll return to drug therapies in the last chapter.

Gene amplification in cancer is only a more extreme version of something that has been a normal part of evolution, probably because it increased the protein coding capacity in vertebrate genomes. Relatively long stretches of

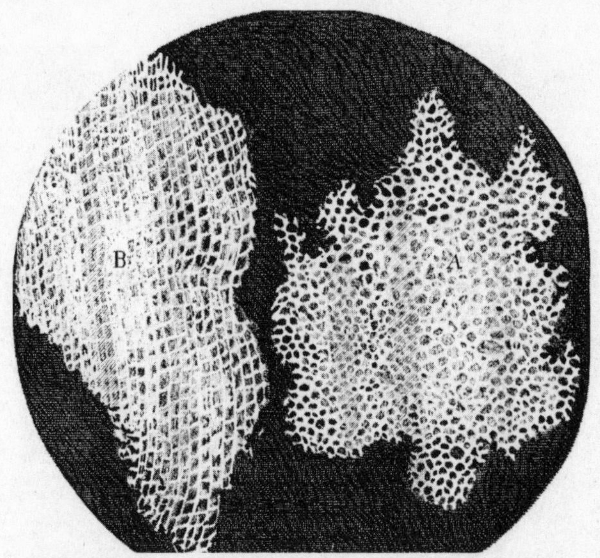

Plate 1: ***Robert Hooke's drawings of the cellular structure of cork in*** **Micrographia** *(1665).*

Plate 2: ***Larger than life*** *(750 million times). This model of the two interwound strands of a DNA double helix is on the path from Addenbrooke's Hospital, Cambridge, the site of the laboratory where DNA was first sequenced, towards The Sanger Centre where part of the complete human DNA sequence was first obtained. The path features the sequence of the BRCA2 gene as 10,257 inlaid bricks in colors representing the four bases A, C, G and T (it therefore encodes a big protein—3419 amino acids). If the pathway covered the whole genome it would encircle the earth 15 times.*

Plate 3: ***Members of the Laboratory of Molecular Biology past and present at Bart Barrell's wedding to Marie-Adèle Rajandream, January 2008.*** *Left: John Walker (who won the 1997 Chemistry Nobel Prize for showing how cells make a useable form of chemical energy). Alan Coulson (behind), Fred Sanger (middle), and Bart Barrell produced the first complete DNA sequence of an organism. Coulson and Barrell subsequently moved to the Sanger Centre to work with John Sulston (right), its first Director, on the project to sequence human DNA. Sanger won the 1958 and 1980 Chemistry Nobel Prizes for his work on protein and DNA sequencing. Sulston shared the 2002 Nobel Prize for his work on animal development.*

Plate 4: ***Sir Peter Medawar (1915–1987)***
Photograph reproduced by kind permission of the President and Fellows of Magdalen College, Oxford.

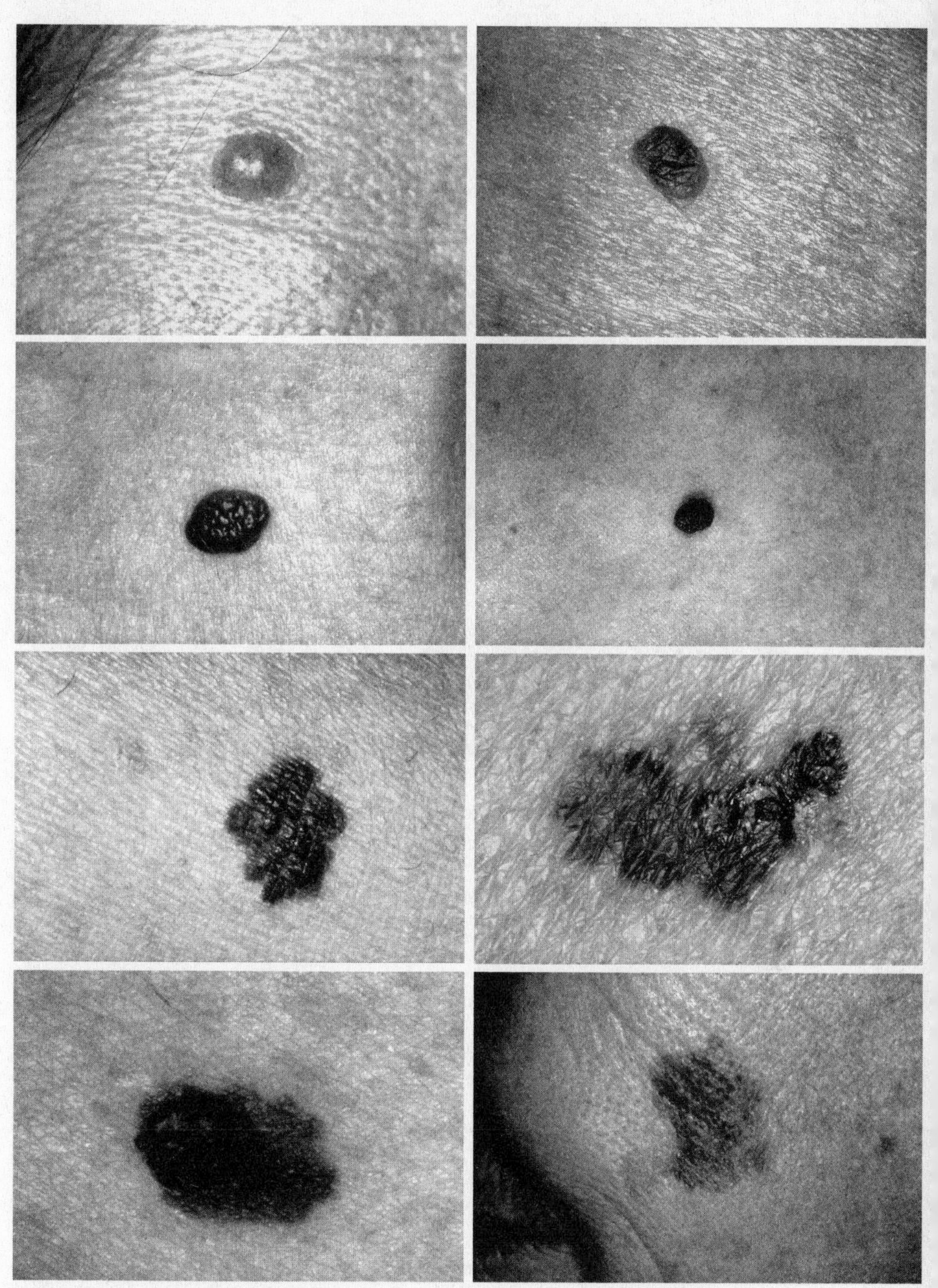

Plate 5: Moles and melanoma. Top four: *moles that are harmless.* ***Bottom four:*** *forms of malignant melanoma. The mnemonic ABCDE describes the early signs of melanoma:* ***A****symmetry,* ***B****orders,* ***C****olor,* ***D****iameter,* ***E****volving. Thus in the top two rows each mole has a fairly uniform color and sharp edges. Melanomas (bottom two rows) vary in color through shades of brown to black and may have regions of white, red, or blue. They have irregular edges, change with time and those shown have grown to a diameter of more than 6 mm (a quarter of an inch). The most aggressive form, nodular melanoma, that may appear where there was no previous lesion, is classified as* ***E****levated (above the skin surface),* ***F****irm (to the touch) and* ***G****rowing. Photos: Skin Cancer Foundation with the permission of The National Cancer Institute.*

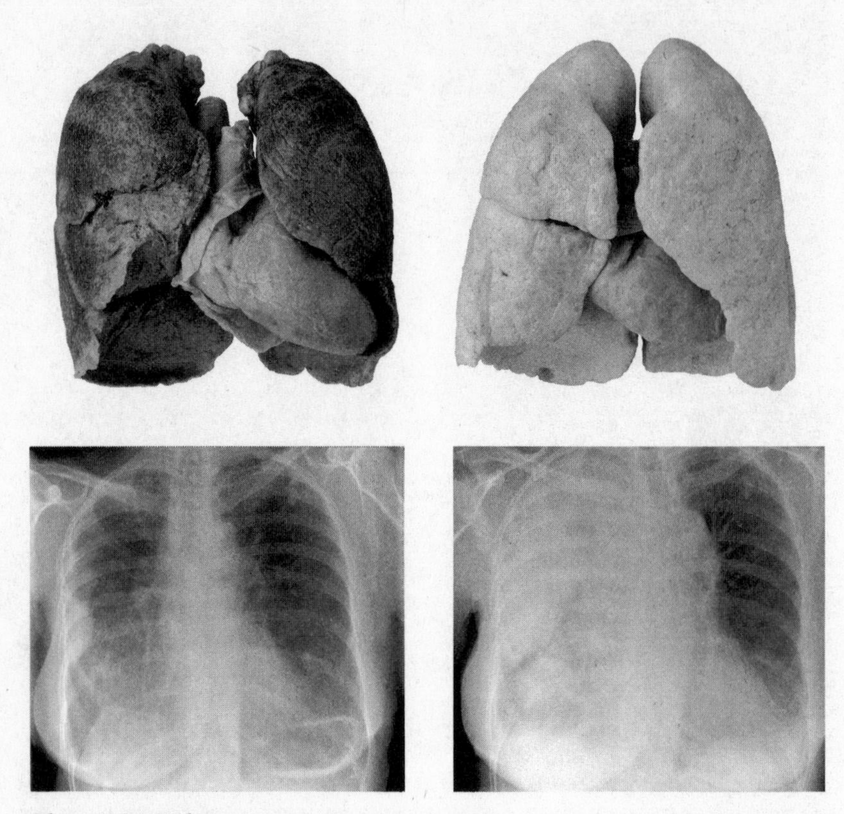

Plate 6. Human lungs.

Top left: *Normal (non-smoker's) lungs.*

Top right: *Smoker's lungs.*

(Gunther von Hagen's Body Worlds, Institute for Plastination, Heidelberg, Germany, www.bodyworlds.com)

Bottom row: *Two chest X-rays of the same female patient showing a tumor at an early stage of development (bottom left) and at an advanced stage (bottom right) (Images kindly contributed by Dr. Ferdia Gallagher, Department of Radiology, Addenbrooke's Hospital, Cambridge and Cancer Research UK Cambridge Research Institute).*

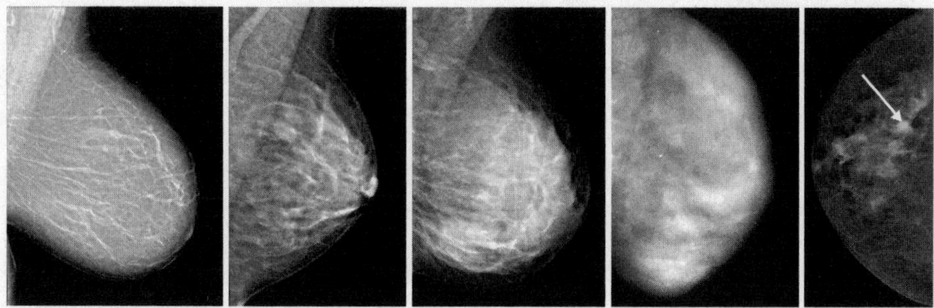

Plate 7. Imaging breast tumors. *Breast tissue is a mixture of fatty tissue, through which X-rays pass relatively easily (dark areas), and denser tissue that absorbs X-rays (white). 7.1, 7.2, 7.3 and 7.4 are normal mammograms of various breast densities. 7.5 shows a tumor (arrow) partially obscured by adjacent dense tissue, showing how difficult it can be to detect a small tumor by mammography (Images kindly contributed by Dr. Peter Britton, Consultant Radiologist, Addenbrooke's Hospital, Cambridge UK).*

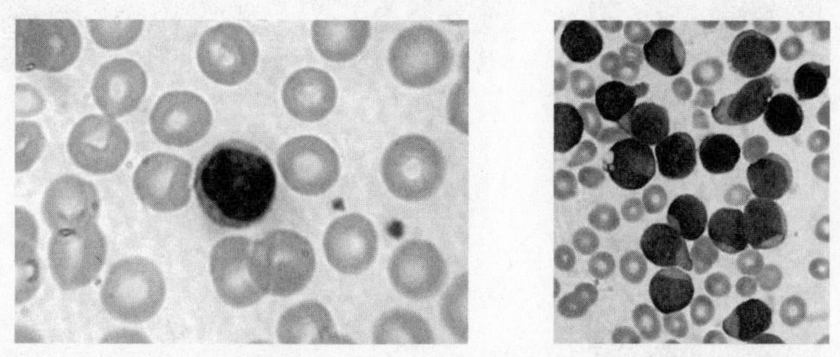

Plate 8. Human blood. *The cells in blood are mainly red blood cells (erythrocytes: flexible biconcave disks with no nucleus about 6–8 μm (millionths of a meter) in diameter) that look like doughnuts.*

Left: *Normal blood comprised mainly of red cells. The large, dark-stained cell is a normal, mature lymphocyte (a white blood cell that makes up 1 percent of blood in a healthy adult).*

Right: *Blood from a patient with acute lymphocytic leukemia (ALL). The dark cells are white blood cells but they are very immature with large nuclei and very little surrounding cytosol (usually called blast cells). (http://commons.wikimedia.org/?title=File:Lymphocyte2.jpg and http://en.wikipedia.org/wiki/File:Acute_leukemia-ALL.jpg)*

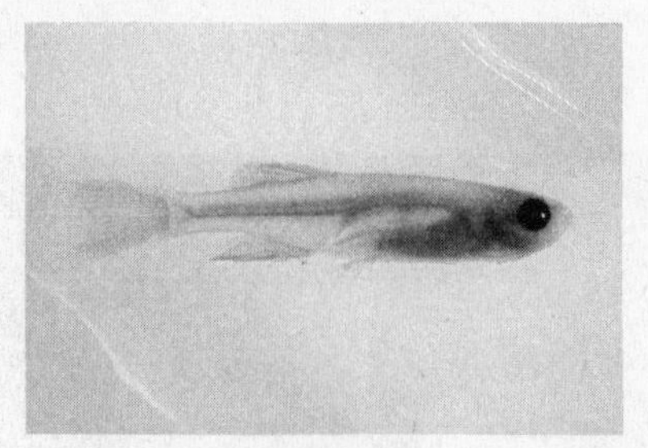

Plate 9. Melanoma in a transgenic zebra fish. *Two mutations prevent these fish from making pigments so that their bodies are largely transparent. The left picture shows one of these "casper" mutants; on the right a melanoma tumor (black mass) has grown rapidly in the abdomen after a small number of tumor cells were injected 14 days earlier. (Courtesy of Richard White, Dana Farber Cancer Institute, Children's Hospital Boston.)*

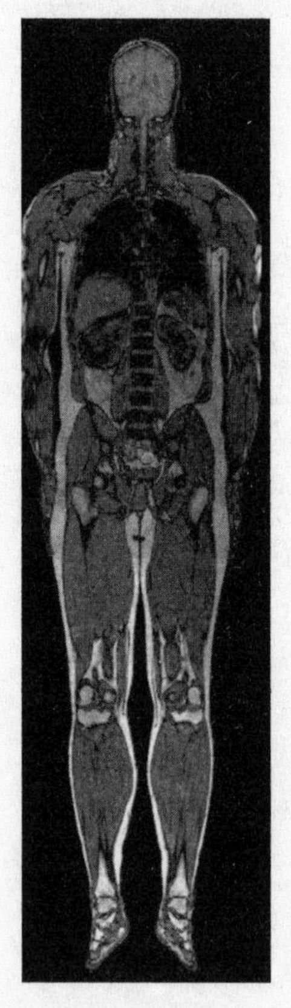

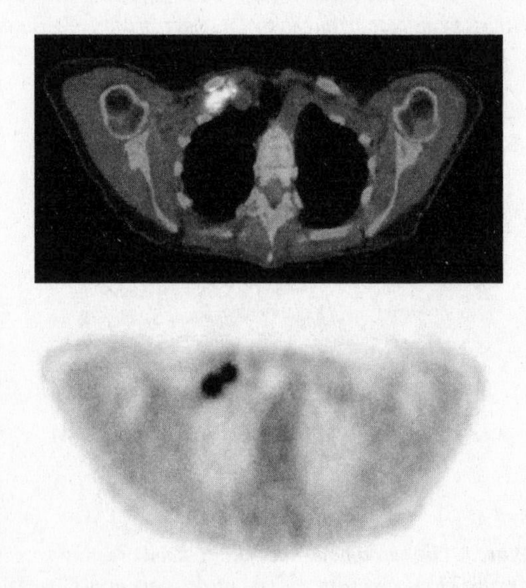

Plate 10. MRI and PET.

Left: *Whole-body MRI imaging showing high soft-tissue contrast (Image kindly contributed by Dr. Peter Börnert, Philips Technologie GmbH).*

Right: *Two fluorine–18-fluorodeoxyglucose PET images of a section through the pelvis.* ***Top:*** *shows mainly bone structures and the more active muscles.*

Bottom: *PET-CT image. These scans are of a patient with metastatic melanoma: the two, small tumors are visible in both but stand out in the lower picture (Images kindly contributed by Dr. John Buscombe, Division of Nuclear Medicine, Addenbrooke's Hospital, Cambridge).*

human DNA are quite frequently duplicated and about 5 percent of our genome is made up of what's called segmental duplications—that is, bits longer than 1,000 bases that are virtually identical in sequence to stretches found elsewhere. More than half the (male) Y chromosome, for example, is made up of this repetitive stuff. The importance of duplication is perhaps best illustrated by the fact that if you compare the whole genomes of any two individuals, on average there will be about eighty genes with a different number of copies—and that difference in "gene dosage" (reflected by the amount of proteins made) makes a significant contribution to the differences between us.

Genes go missing

1: . . . ATGCCGCCCAAAACCCCCCGAAAAACGGCCGCCACCGCC
GCCGCTGCCGCCGCGGAACCCCCGGCACCGC'CGC
CGCCGCCCCCTCCT . . .

2: . . . ____
________________________________ . . .

"This genetics stuff is a doddle, isn't it? Any eejit can see there's no DNA at all in the second line so in that tumor a huge chunk of a chromosome must have got blasted." Right first time! And indeed deletion—the flip side of duplication, i.e., complete loss of a chunk of DNA that includes a gene or genes—is far from a rare event. At first glance it may seem improbable that the loss of a gene, and hence a protein, might promote cancer. However, we referred just now to the examples we've seen of biological systems involving balance between forces pushing in opposite directions. Nowhere is this more true than in the cell cycle that is driven forward (cell proliferation) by the action of kinases but is also negatively regulated—that is, there are proteins that can hold up progression through the cycle (chapter 4). What purpose does that serve? Well, cells should only divide when told to do so by signals from outside. Another major priority is that division should occur *after* the cell has doubled its DNA and shouldn't happen at all if the DNA is damaged in any way.

Accordingly, there are proteins that, in effect, tell the cell to take a break and check that everything is OK. These proteins therefore act as brakes so it's easy to see that if they aren't working, the cell will lose a negative regulator of normal division and become cancer-prone. The genes that encode such proteins are the tumor suppressors that we introduced earlier as the second major group of "cancer genes." These have evolved to control normal cell growth so they're not really "tumor suppressor" genes at all; growth suppressors would be a better name. The critical point, however, is that key members of this growth suppressor family are often knocked out in cancers. This can happen in lots of ways but the mechanics don't matter: the important point is that losing one of these regulators is a green light for cancer to make its move.

SUPPRESSING THE IRREPRESSIBLE

Someone's worked out that if we live to be 140, we'll all get cancer. That should fix the trade in eternal life pills, but the relevant point is that most of us now live to 60 or 70 *without* getting cancer—how does that happen? After all, for virtually the entirety of human evolution, 30 years of age has been the limit—time enough to get a bit of procreation in, which is all that matters (to the species, that is). An absorbing question (at least to me) is how sex came to be so appealing. I mean, if it was as attractive as, say, circuit training and we lived to only 30, we'd have died out. But all that is really secondary to the fact that we now have amazing longevity. Imagine you are related to the Duke of Westminster (who, when I last checked my place in the list, is the richest man in England) and for a wedding present he gave you a motor car. You've used it every day since. Never serviced, the only thing it ever gets is gassed up. To celebrate forty years of bliss you invite your lady wife for a spin in the country and for the first time *ever* your automobile refuses to mobilize! What do you do? (1) Yell down the phone at the duke about giving cheap pressies. (2) Yell at the wife to get under the hood and fix it. Or, (3) sit down with a large brandy and a larger cigar to meditate about the wonders of engineering science that gave you this immensely complicated jalopy that has served you so well for so long? Well, the human body is vastly more complex than any motor, we give it far more abuse, and now we expect it to work—trouble-free, no maintenance—for much longer than thirty or forty years. I know, I know, Americans past a certain age get quite keen on "check-ups," but Brits regard such things as a bit of a stain on their machismo, so they really do rely on Mother Nature. We've seen the numbers that tell us one of the most important things we expect the old girl to keep in check is cancer and, somewhat astonishingly, from the bits and pieces left lying around in our DNA after we got the hang of reproducing and living to thirty, evolution has cobbled together a more than decent cancer defense team.

The first evidence that there was such a thing came from experiments fusing two different types of cell together. This is easier to do than it sounds. A dash of some chemicals can disrupt cell membranes enough to make them stick together (polyethylene glycol—the antifreeze agent in the aforementioned auto's coolant—works well) and, bingo, you have a hybrid cell. When one of the cells was normal and the other was from a tumor, it turned out (almost always) that the hybrid had lost the capacity to form tumors. That was a slightly surprising result at the time, but we now have a ready explanation: the normal cell donates something that the tumor cell has lost (a gene, of course: geneticists call this "complementation"). Knocking out the gene allowed the cells to grow as a tumor—a brake had been released and we've observed that one place where you might expect to find negative regulators is in the machinery of the cell cycle. Two of the most important of all cancer genes do indeed act as brakes on the cell cycle—the retinoblastoma gene (RB1) and P53.

The suggestion that cancers could be caused by losing something was first made by Robert DeMars in 1969. DeMars had been a graduate student of the legendary Salvador Luria (a Nobel Prize winner for his work on viruses that infect bacteria) and worked for many years at the University of Wisconsin, Madison. Alfred Knudson (1971) put his idea on a firm basis mainly by thinking about retinoblastoma, a rare inherited childhood disease in which tumors develop in the eye. The cancer comes in two forms, sporadic and familial. In sporadic retinoblastoma there's no family history and just one tumor develops in one eye. In the inherited form tumors arise in both eyes. Knudson, recalling that we normally have two copies of each gene—each called an allele—suggested that you only got retinoblastoma when both copies of a gene are knocked out. The two forms occur because children either inherit one defective allele and then lose the other (familial) or are born genetically normal but pick up mutations in both genes. Either way up, to get the disease you need to have both copies knocked out.

Because two genetic events are needed, this became the "two-hit model" and a remarkably perceptive bit of thinking it was, given that it was based on no molecular evidence whatsoever. It's turned out to be absolutely correct for the retinoblastoma gene although it was a very long journey from Knudson's idea to the identification and cloning of the gene. We now know that the gene is completely defective in all retinoblastomas but the real importance of RB1 in cancer is that it is also knocked out in a good many other tumors. It's missing, for example, in 20 to 30 percent of lung, breast, and pancreatic tumors. RB1 is therefore the classical model for what's called a "tumor suppressor" gene. *Both* copies of the gene must be inactivated for the tumor to develop (geneticists call this behavior "recessive"—that is, the loss of one copy has no

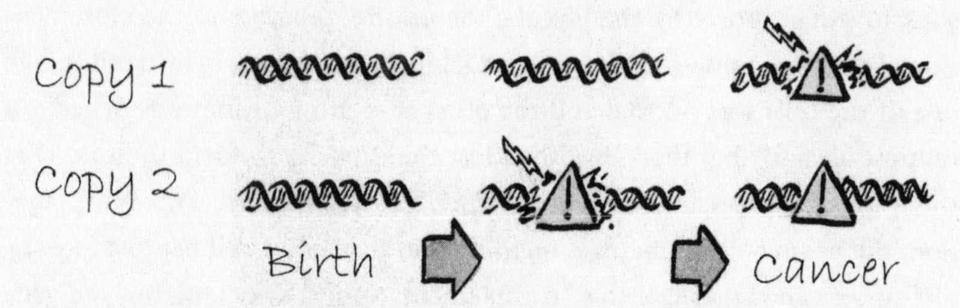

Retinoblastoma. *This cancer occurs only when both copies of a gene are knocked out (!). In one form of the cancer, the individual is born with two normal copies of the gene and both become mutated after birth (they're somatic mutations). In the other (inherited) form, children are born with one defective gene and lose the other copy by somatic mutation. (Mutations acquired by any of the cells of the body except the germ cells (sperm and egg) are somatic mutations: that is, they are not passed on to children).*

effect). Oncogenes by contrast are "dominant," because mutation of just one allele *is* sufficient for an effect to be seen.

What Does the Retinoblastoma Protein Do?

The important thing to grasp is that *all* normal cells have RB1 (two copies), and it's when they're lost that the cell becomes abnormal. Which makes the next question: "What is abnormal?" To answer which we'll look first at normality. The cycle that leads to cell division proceeds in a series of steps and when we walked round it in chapter 4, we saw that the sequence is precisely controlled. The most important step before one cell divides into two is the replication of its DNA. The retinoblastoma protein is critical in this because it controls the cell cycle clock, permitting DNA replication only when the cells are big enough (i.e., when they've passed through the G1 growth phase) and when growth signals are telling the cell to divide. Given its central role in the cell cycle, it will come as no surprise to learn that RB1 is a critical regulator of transcription. RB1 is really a sort of master regulator because it works by associating with a set of conventional transcription factors called the E2F family. RB1 binds to E2Fs, forming a complex that *prevents* transcription of a set of genes. When RB1 releases its hold on the E2Fs, they become free to turn on their target genes. What do the proteins made from these genes do? Well, some are required for DNA replication, some regulate cell cycle progression (MYC is one) and, as if that were not enough, other targets are the enzymes that actually make RNA from DNA—essential for the production of every protein. All of that means that RB1 not only determines whether a cell can replicate its DNA and whether it can carry on round the cell cycle but whether it can function at all.

So that's why RB1 loss is such a big step toward cancer: it blocks cell cycle progression via interaction with the E2F transcription factor family until the time is ripe for the cell to replicate its DNA. At that point, it releases the transcriptional power of the E2Fs. If that control is lost, abnormal duplication of

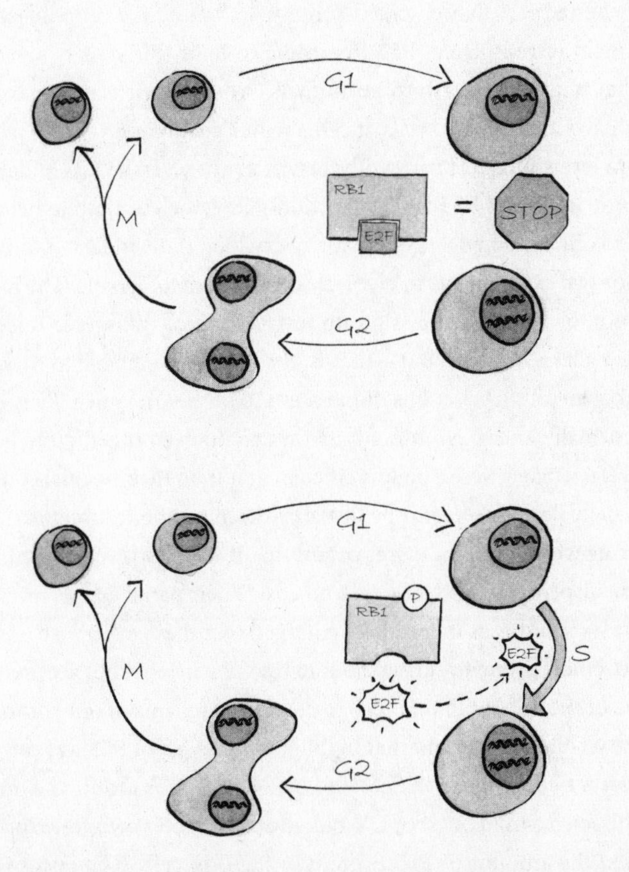

The retinoblastoma protein (RB1) controls the cell cycle clock. *RB1 interacts with E2F proteins to block cell cycle progression from G_1 to S phase. Add phosphate (P) to RB1 and E2F is activated (as a transcription factor—E2F binds to regulatory regions in the genes it controls), and cells progress to S phase.*

DNA ensues and the stage is set for the accumulation of mutations. Only one question remains: How is the release of E2F brought about? RB1 is a target of some of the cyclin-dependent kinases that control cell cycle progression (chapter 4). Under normal circumstances, these control the switch from repression to activation by transferring phosphate to RB1: the change in shape thus caused releases the grip of RB1 on E2Fs.

TUMOR PROTEIN 53

The retinoblastoma protein is a major player in cancer because it normally acts to restrain progression around the cell cycle and hence cell proliferation. Despite the fact that RB1 is *the* model tumor suppressor, it has an even more celebrated counterpart that emerged some years before the retinoblastoma gene and protein were isolated. P53 first came to light thirty years ago through a virus that causes tumors in some animals. When the virus protein responsible was pulled out of infected cells, it came with a cellular protein, P53, stuck to it. Far from joy-riding, its brush with the virus protein is the kiss of death for the virus protein targets P53 for destruction. Nevertheless, despite being known, what P53 did remained a mystery for a very long time. In fact, it took until the 1990s for two experiments to give a clear indication of its role. The first was the generation by Allan Bradley's group in Houston of transgenic mice that had both P53 alleles knocked out—that is, the mice were unable to make any P53 at all. To general surprise, this didn't seem to bother the mice. They developed quite normally and grew into adults that could even breed quite happily, although the females were a bit less efficient at it than their normal counterparts. Fortunately, Bradley's group persisted by keeping the transgenic mice, while wondering why a protein, so important that it was destroyed by a tumor virus, could be dispensed with seemingly no cost. Their patience was rewarded over the next six months as it gradually emerged that these otherwise normal mice were extremely prone to cancer. So much so that after about six months, three quarters of the mice with no P53 had developed tumors of one sort or another.

The second experiment that highlighted the role of P53 was carried out at the University of Dundee in Scotland by David Lane's group, the members of which directed a mild burst of UV radiation onto their own forearms and then measured the amount of P53 protein in the skin cells. The result was amazing: before irradiation, P53 was almost undetectable (as it is in most normal cells), but after a short exposure to sunlight, the amount of P53 had increased dramatically.

Thus was P53 revealed. It isn't necessary for normal cells to grow and divide, indeed it's so unnecessary that normal cells hardly bother to make it. But subject a cell to any kind of stress that damages DNA—even mild UV radiation is very stressful in this respect—and P53 is made in large amounts. This explains why the knock-out mice were fine, but it also explains why they were so susceptible to cancers: all animals are exposed to a wide range of DNA-damaging assaults, mice included, and P53 has evolved to be part of a major mechanism for protecting cells against such assaults. The sunlight experiment showed P53 working overtime to block the mutations that lead to skin cancer. It's a decent protection but, like any defense, if you hit it hard enough for long enough, it will break down, as shown by over a million new cases a year in the US.

How Does P53 Do It?

The cell machinery that detects DNA damage is exquisitely sensitive: it can pick up just one break in the two meters of double-stranded DNA in a cell and, as a result, turn on the P53 gene. There are two ways in which the protein produced could protect cells against DNA damage: (1) stop the cell going round the cell cycle and dividing until the damaged DNA has been fixed, or (2) kill the cell. P53 can do both. It switches on genes, and its first target arrests the cell cycle—recall that the cycle is a balance between kinase-driven progression and various brakes that essentially act as checks to ensure that all is well before the green light for the next stage is given. The brake activated by P53 is a rather promiscuous little protein that blocks the action of several cell cycle kinases, so it's a cyclin-dependent kinase inhibitor.

The first action of P53 is therefore to stop cells from replicating DNA and going on to divide with damaged DNA. This cell cycle arrest buys time—time for the DNA repair system to repair the damaged DNA, whether the mutations have been caused by UV light or by something else. But there are all sorts of reasons why simply giving the cell time might not be enough to deal with DNA damage. When damage occurs, the best anti-cancer tactic is to destroy the cell, and P53's second role is as a suicide activator. If you thought about killing a cell from the inside, so to speak, you might decide that the most obvious target was the central powerhouse, the mitochondrion: stop that and the cell will inevitably die. This is precisely how it has evolved for P53, in that its major targets block mitochondrial function. They do this in the most effective way you could imagine: by making holes in the mitochondrial membrane. This

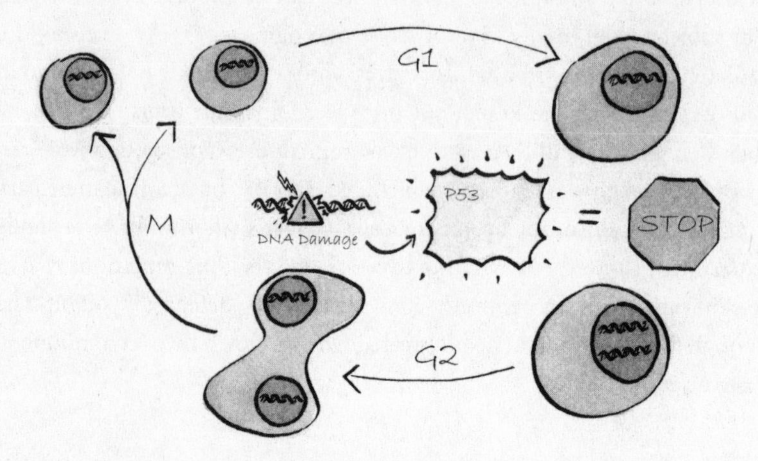

P53 in cell cycle control. *DNA damage raises cellular P53 levels: normal but not mutant P53 can then arrest the cell cycle and promote DNA repair or, ultimately, cause cell death.*

torpedoing effect, set off by P53, is fatal for the cell, for it has switched on the most effective cancer protection you could have: the elimination of a cell that is likely to turn into a tumor precursor.

P53 and Cancer

P53 is therefore central to the way cells respond to stress: it's activated by DNA damage and by lots of other things the cell finds stressful—for example, activated oncogenes or low oxygen levels. P53 then stops division with a back-up strategy of eliminating the cell. This critical role led David Lane to describe P53 with the vivid phrase "guardian of the genome." As cancers almost always involve disruption of the genome, it's unsurprising that P53 is one of the most frequently mutated of all cancer genes. The gene itself is often deleted in human cancers but, in addition, over 6,000 P53 mutations have been detected. All told, over 70 percent of human cancers have a mutation that affects the activity of P53. Mutations are usually not inherited and occur with high frequency in all types of lung cancer, in breast tumors, and in brain tumors, frequently in combination with the activation of oncogenes. However, some unfortunate individuals inherit one defective allele, as can happen in retinoblastoma. This gives rise to a rare disease called Li-Fraumeni syndrome: 50 percent of the carriers develop diverse cancers by thirty years of age, compared with 1 percent in the normal population.

The Double Life of P53

We saw in Chapter 3 how X-rays can be used to "see" proteins, and the structure of P53 attached to DNA gives a very clear picture of how it works to control genes. Rather than binding as a single protein, P53 forms a cluster of four: looking down the axis of double-stranded DNA, it's like a clamp around a pipe or perhaps two pairs of hands gripping a rope. So one could imagine how that cluster might be a target (or part of one) on DNA for the enzyme that makes RNA (that's an RNA polymerase, by the way, because it's an enzyme (an -ase) that makes RNA polymers). That's fine, but how exactly does P53 contact the DNA double helix? We've seen how the coiling of the two strands creates gaps between the strands, the major and minor grooves. These allow proteins to "see" the edges of the bases in DNA, and "fingers" of P53 do indeed slot into regulatory sequences in DNA. Chemicals in tobacco smoke can mutate the contact fingers of P53—a good example of how a carcinogen can promote cancer by having a specific effect on a key tumor suppressor.

The fact that P53 works as a cluster of four molecules means that mutation of just one copy of the P53 gene could give a mixture of mutant and normal forms that might behave abnormally. This can certainly happen—mutation at one allele being sufficient to alter cell behavior—so that means P53 is not following the "two-hit" model. In other words, although P53 is considered to be a member of the tumor suppressor family, it is more complex than the paradigm RB1. This kind of complication isn't confined to P53: quite a few members of the tumor suppressor family can acquire a mutation in one allele that changes the function of the protein, so there's an effect on the cell even though one allele is still normal. Indeed, for P53 at least, it's even more complicated because multiple variants of the protein can be generated, some of which actually behave like oncoproteins rather than tumor suppressors—that is, they actively promote abnormal proliferation.

P53 is therefore remarkable in that, although the normal gene is a potent tumor suppressor, variants can occur that are positively selected for in cancer cells so that the mutant protein is strongly expressed. One effect can be to make cells more invasive—more likely to spread around the body and start secondary tumors.

COLLECTING MUTATIONS

The list of things that can give us cancer has grown so long (UV, X-rays, booze, smoking, certain foods, radon, getting old—OK, that's more than enough) that you might be inclined to wonder how it is that we don't all get cancer at a very early age. Or, put another way, why hasn't cancer finished off the human race? We've already mentioned one reason: the fact that most cancers develop very slowly as we accumulate mutations. One reason progression is so slow is that we have evolved ways of repairing our DNA when it has been damaged. As we've seen, a variety of forms of radiation can disrupt DNA, either directly or indirectly, and some foods contain chemical carcinogens. It's even worse than that because some normal cellular reactions make by-products that can react chemically with DNA to cause mutations—they produce carcinogens by accident, so to speak. Biology is wonderful, but it's far from perfect.

In view of this array of assaults, it's fortunate that we have three lines of defense. The first lies in what we eat. The reason we are so often told to eat fresh fruits and vegetables is that they contain several antioxidants (e.g., vitamins C and E) that confer a degree of protection against DNA damage. The second is that we have evolved a number of proteins that provide a biochemical buffer by reacting with harmful by-products of normal metabolism and converting them into harmless substances.

Nevertheless, despite these protections, it is estimated that the DNA in a normal adult human cell suffers about 20,000 DNA hits–damaging chemical reactions—per day. Many of these might result in mutations *if* the modified, damaged DNA is not repaired. This makes the third and most important protective mechanism a variety of DNA repair processes that correct nearly all of the hits so that, on average, less than one of these gets "fixed"—that is, remains in the DNA to be passed on when the genome is replicated. All this means that mutations accumulate at the rate of one every day or so over a lifetime. Several hundred genes are devoted to various aspects of DNA repair, and it is probably evident that they are an important class of tumor suppressor. Mutations that impair their function will permit the replication of potential cancer cells and, for obvious reasons, mutations in these genes are said to cause genetic instability. The genetic disease xeroderma pigmentosum (XP) is a good example. Sufferers are prone to skin cancers and need to avoid sunlight. Their problem is not with P53, however, which is activated and stops cells dividing, buying the cell time. However, the time bought is of limited use because there is a genetic defect in the system that limits the body's ability to repair DNA damaged by UV light. In addition to XP, a number of other conditions that predispose individuals to cancer are the result of DNA repair defects including Werner syndrome, Bloom syndrome, and Fanconi anemia. Each occurs because of mutations in proteins that signal or enact the repair process. Ataxia telangiectasia, a rare inherited neurodegenerative disease that carries an increased risk of cancers, particularly leukemia and lymphoma, is also often included in this category. It arises from mutations in a protein that normally detects double strand breaks in DNA and then activates P53.

GENETIC ROULETTE

In a normal lifetime, about 0.0003 percent of our DNA gets damaged beyond repair. But the really interesting question is: What happens in a cancer cell? We noted that the cumulative number of mutations can be huge: as many as 100,000 in a cell. Recall that the human genome has 3 billion (3×10^9) base pairs with about 21,000 genes. Encoding these needs only a tiny fraction of our DNA (less than 2 percent). This means that, although the vast majority of mutations occur in intergene regions, there are likely to be about 100 mutations that alter proteins in each cancer cell, i.e., less than 1 percent of genes acquire a mutation. As we've seen, non-coding mutations may alter the behavior of cells, but let's focus on the 100 coding mutations. Of the 100 mutated genes, only a small number are cancer-inducing, that is, they actually *drive* the development of the cancer. With some exceptions, it takes a long time for a critical set of mutations to accumulate in a single cell; that's why most forms of cancer are diseases of old age. Despite the sequence revolution, we still don't know for any cancer precisely what the critical number of hits (i.e., mutations) is, but estimates range from five to fifteen distinct mutations, although fewer may be required for some cancers, in particular leukemias. Whatever the precise number, they make up a set of "driver" mutations sufficient to override the normal controls of cell proliferation.

The five to fifteen distinct "driver" mutations are thus a subgroup of the random mutations that become "fixed" in the genome, and the term "driver" is used to distinguish mutated genes that specifically contribute to cancer development from the "passengers"—all the other mutations that get fixed in the genome but don't do much.

INHERITING CANCER

Ninety percent of these mutations accumulate in the cancer cell during the lifetime of the patient—they're called "somatic mutations." The other 10 percent are set off by inheriting what you might call a starter mutation. That is, it came from either egg or sperm (so it's a germline mutation) and we've already seen that this type of mutation can occur in RB1 and P53. When cancers "run in families" (or show "familial aggregation" as the population geneticists would say), it means that children may inherit a mutation that greatly increases their risk of developing cancer. Because of the risk of being born with a pre-disposing mutation, members of such families have been intensively studied, thereby giving us a picture of the molecular changes that occur as tumors evolve. Two of the most common hereditary cancers are those of the breast and bowel.

Breast cancer is about twice as common in first-degree relatives of women with the disease as it is in the general population (you're a first degree relative if you're someone's parent, offspring, or sibling). About 5 percent of all female breast cancers (men get the disease too but very rarely—about 1 percent of all breast cancers) arise from inherited mutations. In the 1990s it was shown that either of the BRCA1 or BRCA2 genes (Plate 2) can carry mutations that confer a lifetime risk of the disease of over 50 percent, compared with an aver-

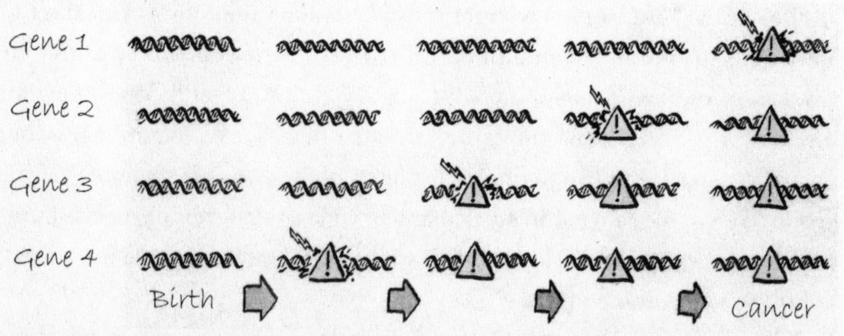

A mutational steeplechase leads to cancer. *Of the tens of thousands of mutations that accumulate over time in a cancer cell, a small number of distinct "drivers" make the cancer develop (four are shown as danger [!] signs). Almost all mutations arise after birth, but about one in every ten cancers start because a person is unfortunate enough to be born with a mutation: they are already one jump ahead and are much more likely to get cancer than those born with a normal set of genes. The rate at which mutations arise is increased by exposure to carcinogens, e.g., in tobacco smoke.*

age risk of about 10 percent. Since then mutations in some other genes have also been shown to increase the risk. However, taken together, they account for only about 25 percent of the risk factors, which means that for three quarters of familial breast cancers the genetic cause is unknown. It seems very likely that the missing factor is the combined effect of lots of DNA variants each making just a small contribution (these are some of the genetic variations that make individuals different, exploited in DNA fingerprinting). As usual, the geneticists have a word for it—"polygenic." So people with the "other" variant have, in effect, a degree of protection against cancer. Twins who are genetically identical have similar risks of developing breast cancer, consistent with the idea that many variants, each having a very small effect, can combine to give a substantial increase in risk.

Cancers of the bowel have perhaps been the most informative for discerning patterns of mutations that drive tumor development. This is particularly true of the inherited disease familial adenomatous polyposis (FAP) in which hundreds of lesions (polyps) form in the colon at an early age, any one of which can, over time, convert into a malignant tumor. It's relatively easy to examine the lining of the bowel and screen individuals from susceptible families. If pre-malignant polyps are detected, they can usually be treated by surgery. Only if further mutations occur can a polyp develop so that it penetrates the underlying membrane, allowing tumor cells to pass into the circulation. The condition has then become malignant and hence life-threatening. The protein that is defective in most cases of FAP normally helps to separate chromosomes as cells divide. The abnormal form causes chromosomes to be torn apart in an unregulated fashion, setting the scene for the accumulation of yet more DNA damage—a characteristic feature of cancers. Studies of families with FAP have shown that very frequently the oncogene RAS is mutated and subsequently P53 is lost—a kind of mutational signature. Other mutations occur but we've already seen why these two genes are often involved in cancer. We'll pick this point up again in the next chapter when we look at how key signal pathways converge on the central players to drive the development of cancer.

We've seen that there are two main groups of "cancer genes"—that is, genes that normally regulate cell behavior but that, as a result of mutation, can become cancer drivers. These are oncogenes (e.g., EGFR, MYC) and tumor suppressor genes (e.g., RB1, P53). Oncogenes have gained function as a result of mutation; tumor suppressors lose activity, often because the gene is completely deleted. The effect of combinations of mutations in these classes of genes is to release cells from the normal controls that regulate proliferation, location and movement. These mutations are almost always acquired as we age, although about one in ten cancers arise from a key mutation being inherited. No two people have exactly the same DNA code (except for identical twins) and subtle differences between our genomes can affect our cancer risk. Armed with all this knowledge about genes, proteins, and mutations we'll now look at how they can affect the way the world talks to cells and the overall results of signaling going wrong in cells.

6

A CASE OF CORRUPTION

SCIENCE IS A FUNNY BUSINESS. IN THE END IT'S THE MINUTE DETAILS that are important (think of *The Simpsons*' resident Dr. Nick, waving a scalpel somewhere over your appendicitis, chanting to the operating theater, "south of the stomach, north of the knee, slice it open and let's see"). Nevertheless, vital though it is to know details, cells turned out to be amazingly simple when we looked at their workings without worrying about how many bits there were or what they were called. Indeed, one way of judging scientists (and physicians) is by how good they are at simple explanations of complicated things. So haven't we done well so far?! Fortunately, you don't have time to answer that because we must now put some of our mutations into context to see how they affect signal pathways—the info-flow of the cell. We've hinted at there being lots of these cellular phone lines, and it's easy to make a rather frightening subway map if you put them together. So we won't do that—we'll just take one main pathway to get the general idea, and that will lead us into a "central axis," the real heart of our cancer defenses. It's called the MAPK pathway—it really doesn't matter what the acronym stands for—it's just a top example because it so often gets corrupted in cancers. Even better, we've already met the key pathway proteins in the last chapter when we used them to show how mutations affect specific proteins—cunning or what?!

One small word of warning: there are a few new players and at least one quirky cell-type trick, grappling with which limits the number of funnies in the next few pages—but stay with it. The good news is this is the shortest bit of the story, it's downhill all the way afterwards, and I promise that any brow-furrowing will be worth it. We'll retrace "A talks to B talks to C," the path we took to show—in principle, as the boffins like to say—how the world talks to cells, specifically to the bit that matters—the control center of the nucleus. To see exactly what happens when a normal cell turns into a cancer cell, we need to replace the unknown by the specific—so we'll swap: Receptor → A → B → C → D → E → F → G for the players in the MAPK pathway: EGFR → GRB → SOS → RAS → RAF → MAPKK → MAPK → Nucleus.

We know from the anonymous pathway how phosphate groups on activated receptors act as launch pads for signal pathways, and we know that there can be quite a few of these on one receptor—a kind of decorated scaffold. Now the cluster of pads has been created by the action of the growth factor EGF on its receptor (EGFR), and that's given us quite a choice of pathways down which we could wander. Sad to report and unlike strolls in the country, protein trails are rather repetitive (detail apart!); we'll explore only the one that rejoices in the name of the MAPK pathway. The first steps involve two little linkers (GRB and SOS) that attract the RAS "switch," flipped "on" by swapping G2P for G3P (chapter 5). Three subsequent protein associations, RAS → RAF → MAPKK

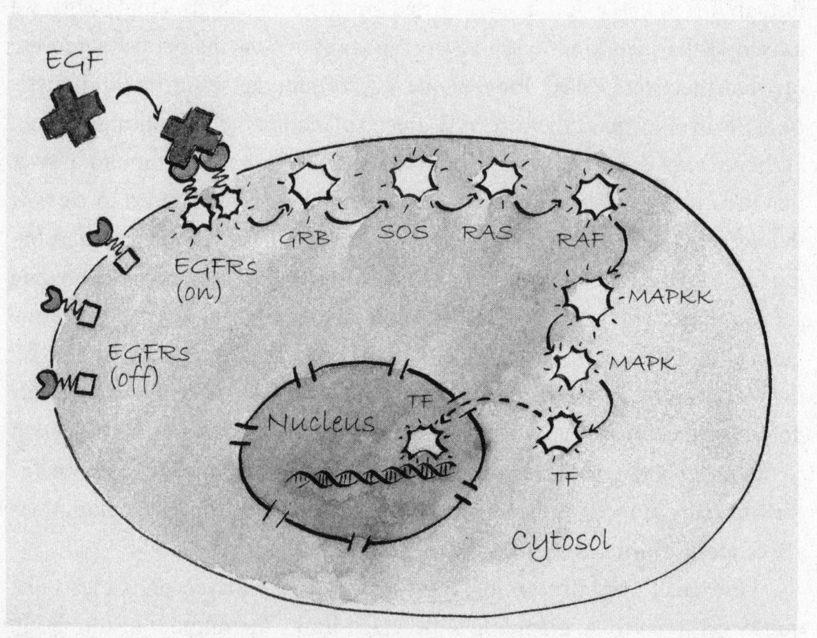

Message to nucleus: Getting at DNA. *In this example the hormone (often called a growth factor) is epidermal growth factor (EGF), which binds to its receptor (EGFR) and switches on a relay of proteins called the RAS-MAPK pathway to carry its signal from the membrane to the nucleus (MAPK =* mitogen-activated protein kinase: *"mitogen" is another name for growth factor). TF = transcription factor.*

→ MAPK, cascade to the nucleus, activating transcription factors (by phosphorylation). In one sense, these steps are simply repetitions of the first three: when one protein contacts another, it causes a shape change so that the second protein talks to a third and hence relays the message. Here, though, there's an extra level of signaling subtlety: at each of these steps, the proteins switched on are enzymes. Once activated, they'll go on phosphorylating their target until such time as something specifically switches them off. So these signaling proteins are yet more examples of molecular switches: they are switched on by the addition of a phosphate group, and as long as that group remains attached, they no longer require contact with their upstream activator to be "on." The only way to switch them off is to remove the activating phosphate group: that, of course, requires another member(s) of the phosphatase family that we ran into in the cell cycle. So RAF, MAPKK, and MAPK form an enzyme cascade. Each adds an amplification step to the signal pathway. In this way, a signal from outside the cell reaches the nucleus. There it can contribute to changing the pattern of gene expression—and hence the behavior of the cell—and, unless it's a steroid hormone, it has done so without the messenger's entering the cell.

NATURAL CONSERVATION

The importance of the MAPK pathway is underlined by the fact that closely similar versions are present in all animals. This theme of conservation runs throughout biology and is reflected by DNA and protein sequences that indicate the common ancestry both of individual genes, as for hemoglobin and myoglobin (chapter 3), and of species (chapter 4). Not only is the MAPK cascade conserved across all organisms as a central mechanism for signal transduction but there are also several closely related MAPK pathways that respond to different types of signal (hormones, stress signals, etc.). They're closely related because the individual proteins are very similar (similar in sequence, similar in shape—just slightly tweaked to do their specific jobs).

THE MAPK PATHWAY GOES ASTRAY IN CANCERS

It is not, of course, coincidence that we've already met two key members of this pathway as examples of oncogene-generating mutations—EGFR and RAS. For RAS we saw in chapter 5 that mutation blocks the switch: it can't convert G3P to G2P so that any signal pathway containing "oncogenic" RAS is permanently activated. The EGFR is present on the surface of most types of cell and, as with our generalized receptor, its activation turns on signaling pathways regulating proliferation and survival. It is perhaps not surprising that the EGFR is frequently abnormal in human cancers, and this can happen in all sorts of ways, as we've seen—by amplification (the receptor is normal, there's just lots more of it doing the signaling), by losing the bit messengers bind to (which turns it on regardless of whether it should be turned on or not), and also by a range of other mutations. By all manner of changes, therefore, can EGFR signaling become hyperactivated.

The central role of the MAPK pathway in cell signaling explains why mutations in receptors like the EGFR and in RAS play such important roles in cancer. It's slightly odd that, although oncogenic forms of receptors and RAS are common in human cancers, mutations in other parts of the MAPK pathway are relatively rare. One exception to this is the RAF family member BRAF that is mutated in most melanomas. RAF proteins are kinases, as we have seen, so it will be no surprise that the mutation in BRAF switches it on as a "phosphorylator." All of these types of mutation have similar effects: through the permanent activation of a critical regulator of cell proliferation, the cells have become largely immune to environmental cues and taken a big step toward becoming tumors.

These general features raise the question of why some signaling genes are more prone to mutation than others. In part, the answer lies in differences between tissues, and we've seen that mutation of BRAF is very common in melanocytes when they develop into melanomas, although RAF family mutations are generally rare. Similarly, almost all pancreatic tumors have a RAS mutation although these only occur in about 20 percent of other cancers. You might guess, therefore, that some types of cancer are identifiable by specific mutations, for example, the amplification of ERBB2 in breast tumors. Nonetheless, cancers are generally characterized more by the diversity of mutational patterns than by uniformity.

A METRO MAP OF CANCER

We've now seen from a few key examples how mutations create "cancer genes" and how these change the behavior of a cell from normal to cancerous. We've seen that there are several hundred such genes, so you might guess that if we assembled all of them in their pathways, it would look like a subway map, as we noted in chapter 4—even though the way the individual

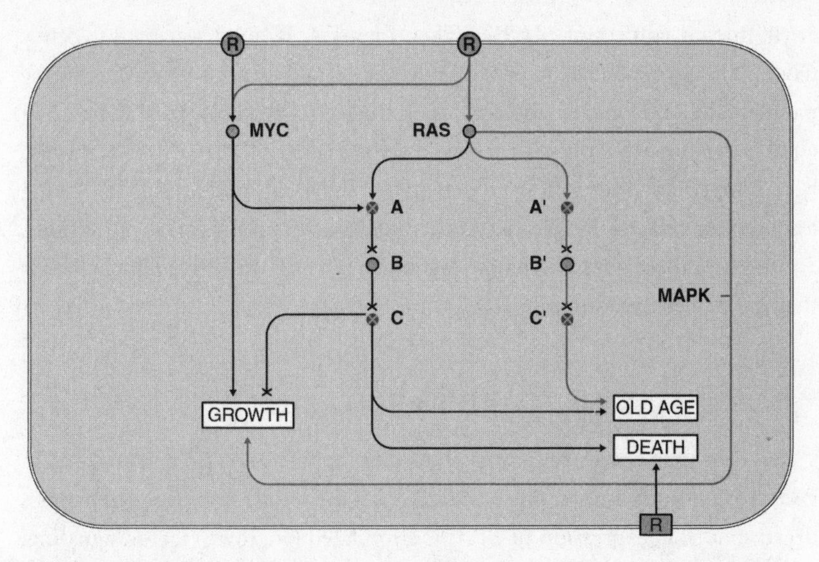

Metro map of cancer pathways and cellular responses. *In addition to growth factor messengers activating receptors (R), lots of other types of receptor can turn on signal pathways (e.g., receptors sensing cell-cell contact or transmitting death signals). From all of this two central regulators emerge, MYC and RAS. They regulate a central axis of two halves that are mirror images in how they work: each feature three key proteins (A, B, and C and A|Prime|, B|Prime|, and C|Prime|). The stations marked with a cross are tumor suppressors; an X on the line means a block. These pathways impact on the three main options a cell has, to grow (and proliferate—which also means to supply itself with nutrients, etc.), to enter senescence (old age), or to die (apoptosis).*

pathways work is pretty simple. So let's not do that but instead represent just a few pathways and include only major players in the cancer game—but for clarity we'll draw it in the style of Harry Beck (who created the London Underground Tube map in 1931) so all we have to do is find our way round a new (small!) town.

At first glance, unfolding our Metro map reveals the expected. There are several lines coming in from the suburbs, with receptors at the start of each line. It's true there are only three ultimate termini—Growth, Old Age, and Death—but that's not really a surprise either for we know this is not really a railway map but an allegory of life. Growth, of course, means tumor cells getting bigger and then dividing as well as the supporting stuff—making new blood vessels, adjusting metabolism, etc., that we know all about. Old Age is what the scientists call senescence—a sort of vegetative state that can protect against cancer—and Death (or apoptosis, as we more delicately call it) is the best anti-cancer ploy we have. But why are there only two named stations (well, three if you count MAPK)? Then there's A, B, and C—why don't they have proper names? They're presumably related to the ones with dashes on the parallel line. And what do the crosses on the tracks mean? At least there's one good thing: the two names we *are* given are old friends by now: RAS and MYC have popped up time and again—as major signaling molecules, MYC being essential for cells to divide, and as examples of important cancer mutations. So we know these guys are major league players, and we'd have felt cheated if they hadn't been on our map.

MYC the Master

We've seen that cancer cells very often lose control of MYC and make it in excessive amounts—up to one hundred times the usual level. That can happen because normal expression of MYC is stimulated by growth factor signaling, and we've seen how often those pathways become corrupted in tumor development—though sometimes MYC itself is mutated. Because MYC must be expressed for cells to divide, it's not surprising that its over-expression is so common in cancers. It's just what the doctor didn't order, as you might say. MYC is indeed one of the most prominent "cancer genes" and is a very good example of what has been called "oncogene addiction," namely, the dependence of continued cancer growth on one specific activity. We might recall too the reason that MYC is so potent—as a transcription factor it can control such a huge number of genes. That's why it turns on not only many of the bits and pieces of the cell cycle but also the supporting acts of tumor growth—nutrient uptake, new blood vessel growth, etc. MYC is closely matched by RAS in terms of the range of effects it can have on cells. We've seen the importance of RAS in controlling the key MAPK growth pathway, but it's also a point of divergence for several other signals that, in particular, regulate important steps in cell division.

So we're not surprised to see MYC and RAS stand out on our map. But there's something striking about where the lines go after them: to the A to C lines. It looks as though that's where the really big hitters are—a sort of central axis right in the middle of town. We're not saying that more distant bits—cellular equivalents of Pelham Bay Park or Epping—aren't important: of course they are, dear residents—we've shown that by taking a receptor (EGFR) as another of our key examples of cancer genes. It's just that Grand Central in New York and Bank or Oxford Circus in London are where the action really is.

The Central Axis

In our tour of cancer highlights we've seen that the message is, as lyricist Johnny Mercer put it, to "accentuate the positive, eliminate the negative"—that is, mutations make cells grow and divide by pressing accelerators and knocking out brakes. The pathways we've looked at have been smooth progressions: A turns on B, which turns on C, etc., sometimes even with an in-built amplifier. The central axis, however, is something new, as indicated by the crosses on the tracks. The new idea is negative regulation: instead of A switching on B, it does the opposite. And B, to make things even more complicated, talks to C but turns it off, too! So we have two negatives in one short pathway. What's going on?

Both A and C (and also A' and C') are tumor suppressors. B (and B') are oncoproteins, but they can knock out C (or C'). B literally does this for C by tagging it for breakdown. B' is less drastic, merely phosphorylating C', thereby releasing a latent cell growth–promoting activity.

These are amazingly subtle examples of control. As the map shows, their outputs are aimed directly at the heart of life: the cell division cycle, senescence, and cell death—presumably why evolution has spent a fair bit of time coming up with such clever interlocking systems. By now, of course, you've guessed at least who C and C' are: P53 and RB1. The regulator of P53 (B) is a protein called MDM2: when it binds to P53, MDM2 stops it from behaving as a transcription factor and hence blocks its key actions: P53 can no longer arrest the cell cycle and cause cell death. Because MDM2 can also destroy P53, it

A	✓	✗	✗	✓	✓
⊥					
B	✓	✓	✗	✓	✗
⊥					
C	✓	✓	✗	✗	✓
Tumor ?	No	Yes	No	Yes	No

Double negative signaling. Left: *A, B, and C interact but, in contrast to the MAPK pathway, A blocks the action of B, and B inhibits C. The rationale is that A and C are tumor suppressors, and B is an oncoprotein. The ticks indicate normal protein. When all are present, the effect is tumor block ("No"). Knock out A or C, and tumors grow. Loss of B removes the suppressor of a tumor suppressor and tumor growth remains blocked.*

delivers a double blow. You couldn't think of a better way of removing a critical tumor suppressor: first stop it from doing its job, then get rid of it altogether. Being able to neutralize P53 makes MDM2 an oncoprotein, and some human tumors make excessive amounts of it. The notion that slightly wonky control of one gene could knock out a key protector would be a bit worrying were it not for the second layer of negativity (that's A). MDM2 binds to P53, but a third protein, ARF, binds to MDM2—even when it's attached to P53—and when it does so, MDM2 can no longer inactivate P53. So that makes ARF a tumor suppressor—it's the protector of P53, a kind of "super-controller" of this pathway—and this key role makes it a massive target for cancer-promoting mutations that knock out its function. It is, in fact, one of the most frequently mutated of all genes in human cancers.

In the mirror image pathway, the regulator of RB1 corresponding to ARF is INK4. INK4A inhibits the kinase that phosphorylates RB1 (CDK) and releases the activity of E2F as a transcription factor, in effect switching on the cell cycle—so in blocking this step, INK4A becomes a powerful cell-cycle brake. Although INK4A works in a different way, its function precisely mimics that of ARF. They are both master tumor suppressors and INK4A too, just like ARF, is inactivated in many cancers. MDM2 and P53 bring to mind a bear that's caught a salmon: as it climbs out of the river, the fish (P53) seems fine . . . it can still do almost everything fish do: it's only when the bear snaps its jaws that it's

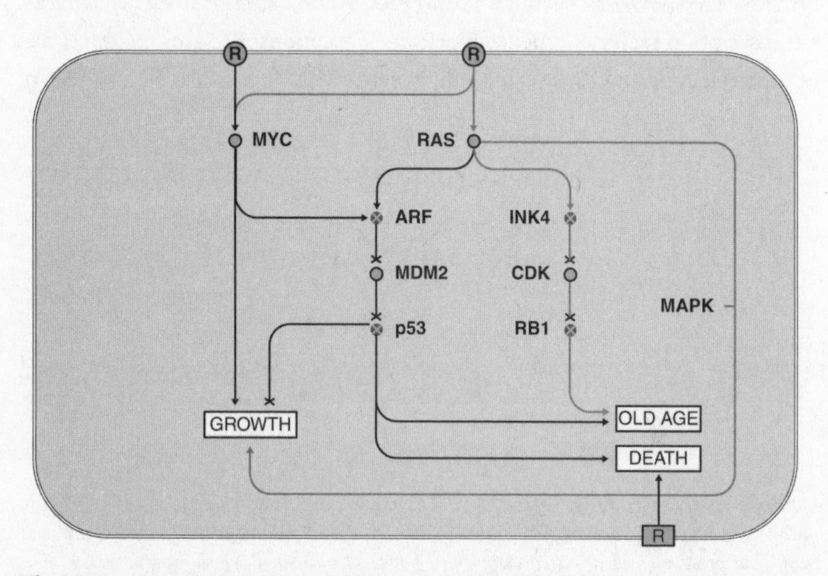

The Metro map: naming names. *INK4 is an* In*hibitor of cyclin-dependent* K*inases, the key cell cycle regulators we met in Chapter 4.*

in real trouble. ARF, by contrast, is a crafty older bear who's waited on the bank and now taps junior on the shoulder, telling him to release P53.

GETTING OLD AND BLOCKING CANCER

We saw earlier that primary cells grown in the lab gradually expire into old age, a state referred to as senescence. Only in the last few years has senescence been recognized as a cancer protection mechanism, and here too P53 and RB1 play central roles. Abnormal stress-like signals activate either the ARF-P53 or the INK4A-RB1 pathway to force cells into senescence and prevent cancer progression. In the end, as the disease develops, senescence is unblocked and proliferation activated, usually by loss of P53.

Thus is revealed the central importance of ARF, INK4A, P53, and RB1 in our cancer defenses. As long as they are present and working properly, we have two very effective strategies: apoptosis and senescence. Knock out even one and tumor development is almost inevitable. So there are two ways of looking at this: (1) most of us don't die from cancer, and most who do are getting on a bit, so this randomly evolved protection system works rather well; or (2) because our protection system is the product of unintelligent design, we ought to be able to do better. Should we consider re-programming ourselves by, for example, boosting our INK4A or P53 stock? Well, mice suggest that we should—that is, transgenic mice engineered to express three normal copies of either the P53 or the INK4 genes (rather than the usual two) are significantly protected from cancer. Given that P53 can drive senescence, you might wonder whether mice with an extra P53 gene might age prematurely, but, so long as the extra P53 is under normal control, they don't and seem otherwise fine. What's more, in mouse tumors generated by knocking out P53, the restoration of P53 expression can cause tumors to shrink and become senescent. The tumor cells actually get broken down with the result that the tumors disappear. All this suggests that being able to re-program our tumor suppressors may be a viable therapy but, given what we already know about the complexity of P53 in particular, we can be sure that this approach will not be straightforward.

HOW CELLS CHANGE AS TUMORS DEVELOP

In the end, of course, the critical question becomes: When the elaborate defense systems that have evolved fail, what is the effect on the cancer cell as a whole of all the cell-signaling upheavals we've talked about? The Metro map really represents a summary of how signaling pathways impact on gene expression in the nucleus. In cancer cells the result is that they have usually (1) lost the ability to go into senescence or to die, (2) found a way of surviving indefinitely, (3) changed their pattern of cell surface and released proteins so that they stick to different surfaces and migrate to where they shouldn't, (4) become able to induce new blood vessels, and (5) adjusted their metabolism to support their new lifestyle.

In short, the effects are to coordinate the modulation of all cellular responses to make a cancer cell able to thrive independently of normal controls, and to have the capacity to colonize new locations, ultimately the most lethal of all the properties of a tumor cell. There are, of course, other genes and other pathways that can be corrupted and help to drive cancer. One reason cancer is complicated is that there are many ways of disrupting normal cell control, and any bit can go off the rails. However, the small number of examples we've looked at tell us all we need to know before we turn to the really exciting part: What actually happens to the cells that suffer such perturbations as they turn into cancers?

PART III

CANCER CELLS AND WHOLE GENOMES

7

TUMORS

WHAT ARE THEY?

ALTHOUGH HUNDREDS OF GENES MAY BE MUTATED IN CANCERS, A relatively small group act as "drivers" in the development of individual tumors. Their overall effect is the corruption of signaling; we now turn to how these changes in the underlying molecular biology affect the way the cells themselves behave. We begin by distinguishing benign (harmless) from malignant (nasty) tumors. The latter evade restraining signals by continuing to proliferate, declining the suicide option, adjusting what they eat and how they use it to adapt to their abnormal environment, and then subverting host cells around them so that they support the tumor rather than destroying it. Finally, the most malevolent property of all is that these aberrant cells begin spreading to distant sites, a process that is only just beginning to be unraveled at the molecular level and remains essentially untreatable—the greatest cancer challenge.

There are about two hundred different types of cancer, but they all have in common abnormal cell growth and the potential to spread from their original primary site to other parts of the body. The act of dispersing from the primary site to form secondary tumors is called metastasis and it's important because it's the main cause of cancer deaths. "Tumor" is the Latin for "swelling," i.e., an abnormal growth, which in current English tends to be used interchangeably with "cancer." Doctors quite like "neoplasm" (new growth) but all three words have much the same meaning. The only question that really matters is whether the cancer is metastatic: Has it acquired the capacity to invade its surroundings, the first step in spreading to secondary sites? To do this, tumors have to destroy other cells so they can move around and, in particular, they have to chew through the walls of the circulatory systems. Once this destruction has been achieved, a tumor cell can be carried in the bloodstream to other locations: it has become malignant. The implication, of course, is that there are tumors that are *not* malignant. These are called benign tumors, and cancers may therefore be divided into two major groups: malignant and benign. Before we turn to malignant cancers, a word about benign neoplasms, not least because practically everyone has some.

BENIGN TUMORS

The critical thing about benign tumors is that they're not malignant—that is, the cells don't invade surrounding tissues and, therefore, they do not spread. They're usually surrounded by a membrane, a sort of sac that helps to prevent them from spreading (they are "encapsulated"). Benign tumors can arise in any tissue—though they're most common in the lungs—and in general they are fairly harmless. Nevertheless, although slow growing, they can reach a considerable size (as big as a grapefruit) and if they compress other tissues (e.g., blood vessels or the brain), they can have serious effects that require surgical treatment. In addition, some benign tumors have harmful, indirect effects when they grow in tissues that make hormones. These include places like adrenal glands or the thyroid, and it's easy to see that if they become enlarged, you'll probably have abnormal levels of the hormones normally made by those glands. In these cases the tumors form from cells of the tissue, which is why they're often symptom-free and only detected by chance (say from a blood test). Just for the record, benign tumors of this type are known as hamartomas. When cells from a tissue form a benign tumor somewhere else, they're choristomas.

WARTS AND ALL

Most of us probably try quite hard not to think about cancer, at least in a personal sense. However, almost all of us will have given it a moment's thought when contemplating the various birthmarks, moles, and warts that adorn most human bodies. The thought will have come as two questions: Are these cancers? What should I do about them? To which the answers are "no" and "nothing"—almost always. In a strictly technical sense they are indeed "neoplasms" because they are an abnormal growth of skin, but the best thing to do is simply regard them as a blemish or, if you like, an adornment—perhaps most famously accomplished by Marilyn Monroe and Mikhail Gorbachev. Some birthmarks, for example "strawberry marks" that generally occur on the face, gradually disappear of their own accord. For marks that do not and are felt to be disfiguring, it may be possible to reduce their prominence by laser treatment.

There is, however, just one word of warning hidden in the "almost always" above. The medical fraternity refer to birthmarks as nevi, the most common nevus being composed of melanocytes (melanin-producing cells) in the outer layer of the skin. Thus moles are benign tumors formed of clusters of pigmented skin cells. The only real problem with moles is that just occasionally one of them may turn nasty and develop into a fully malignant tumor. There are two reasons why even this should not keep you awake at night. First, that event usually needs some help from you to trigger the nasty turn—something that large numbers of us provide by lying in the sun without any protection. Second, because these are skin growths, it should be easy to notice any change in their behavior and have this change checked out by a physician. This is why you are advised to use sun creams and to take action if any of your moles change appearance: get bigger, blacker, or itchier or start bleeding. Cysts are another distinct abnormality. These are not benign tumors but closed sacs of cells containing liquid or semisolid material; commonly referred to as sebaceous cysts, they can be removed by surgery.

Warts have much in common with birthmarks and moles, but there is one big difference: warts are caused by viral infection. For that reason we aren't born with them and indeed they're rare in babies. Nevertheless, most of us get them at some point, often before we are twenty. A recent survey of children in the UK revealed that almost all of them had warts of some description. Before that deters you from giving your offspring a hug, we should point out that almost all warts are harmless and disappear of their own accord, though some take years to do so. Warts are rough lumps of skin that commonly arise on the hands or feet or in the anogenital area—that's the area between the anus and the base of the penis or the vagina. Palmar warts occur on the palm of the hand, plantar warts, otherwise known as verrucas (*verruca plantaris*), on the soles of the feet. Warts are quite different from athlete's foot (*tinea pedis*), a fungal infection of the skin between the toes, but both may be picked up in showers and bathrooms or as a result of sharing towels or clothing as well as, of course, through direct contact with infected skin.

Warts are contagious because of their viral cause, the virus in question being human papillomavirus (HPV), one of the DNA viruses we met in chapter 3. There are more than 100 types of HPV, different types causing different wart variants. HPV produces warts because it causes cells in the outer layer of skin (epidermis) to make excessive amounts of a protein called keratin. Once infected, you can't get rid of HPV. Nevertheless, most warts can be treated either chemically or by freezing (cryosurgery), burning (cauterization), or laser treatment.

So the vast majority of HPVs may have some rather unappealing effects but they're not life-threatening. Unfortunately, fifteen of their considerable number are; of these the most important are types 16 and 18. These cause about 70 percent of cervical cancers and a variety of other anogenital tumors as well as some cancers of the mouth, voice box, windpipe, and lung. One of the critical actions of the tumor-causing HPVs is the inactivation of P53. We have seen the importance of this protein as a transcription factor and how mutations can cause loss of its cancer-protective activity. The virus knocks out P53 by making a protein that binds to it and then acts as a signal to tell the cell to break down P53, thus removing a major proliferation brake.

CROSSING OVER: THE BOUNDARY BETWEEN BENIGN AND MALIGNANT

Most benign tumors are therefore harmless—that is, they never become malignant—but there are some that just occasionally do, and the boundary between them is not very clear. Can we clarify things by looking at the underlying molecular biology? If cancers arise because of mutations in genes, maybe benign tumors are mutation-free and simply arise because of a local imbalance in growth factors. There's much less information available for benign lesions compared with tumors, but a quick glance dashes any hopes of a simple answer.

Some benign tumors never become malignant yet carry mutations in genes that in other tissues are associated with malignancy. On the other hand, there are tissues in which mutations occur in both benign and malignant tumor forms, albeit sometimes with quite different frequency. Therefore even knowing something about the mutational state of genes doesn't enable us to say for certain whether a tumor will become malignant if left to its own devices.

We're stuck with a typical cancer problem. The distinction between benign and malignant tumors is crucial: one of them can kill you. However, even with the power of modern molecular biology, science cannot as yet identify what it is that converts a relatively harmless form of abnormal growth into the fatal variety. The best bet in terms of identifying a suspicious growth remains getting a skilled pathologist to look at a sample of the tumor via biopsy.

CANCER CLASSES

Having made the distinction between benign neoplasms and tumors, let us turn now to what people usually mean when they speak of cancers—that is, abnormal growths that have the capacity to become malignant.

Cancers can be divided into two broad classes: solid tumors and liquid tumors. Most human and animal cancers are solid tumors. Approximately 85 percent are carcinomas—malignant tumors of epithelial cells. Of the four major types of tissue found in the body, epithelia are by far the most prolific. Epithelial cells line the surfaces of the body: the outer layer of skin is a form of epithelium, and epithelial cells are what you scrape off the inside of your cheek. Glandular epithelial cells line the intestine and cover the major glands; cancers that arise in this type of epithelia are called adenocarcinomas. Carcinoma *in situ* is a pre-malignant change that happens in many cancers in which cells proliferate abnormally within their normal location: the epithelial cells show many malignant changes but have not invaded the underlying tissue. Ductal carcinoma *in situ* (DCIS) is one of the two most common forms of breast cancer, characterized by abnormal proliferation in the wall of the milk ducts. It carries a risk of developing into the invasive ductal carcinoma (IDC) form in which the cells are malignant.

Most other cancers are sarcomas, tumors arising in the tissues that essentially hold us together (connective tissue, bone, cartilage, muscle, fat, blood vessels), and they are often highly malignant. However, they are also relatively rare and they are not thought to have a pre-malignant (*in situ*) phase.

The remaining 3 percent of human cancers are leukemias and lymphomas arising from the abnormal proliferation of white blood cells or, very rarely, of cells that develop into red blood cells. The word "leukemia" comes from the Greek for "white" (*leukos*) and "blood" (*haima*): collectively, these diseases are called hematological neoplasms. White blood cells are sometimes called leukocytes but, as that term covers all white cells including lymphocytes, it's apt to be a bit confusing. Lymphomas are cancers of lymphocytes, the cells of the immune system that fall into the two main classes of T cells and B cells. The two major divisions of this cancer are Hodgkin's disease (Hodgkin's lymphoma) and non-Hodgkin's lymphoma.

WHAT MAKES A CANCER CELL A CANCER CELL?

In the first year of the new millennium, two American scientists, Douglas Hanahan and Robert Weinberg, published a review that summarized the features that distinguish a cancer cell from its normal counterpart. Not the least notable aspect of their effort is that it has been very widely accepted in a research community not known for harmony and uniformity of view. You may well be wondering why it's taken over three thousand years to come up with a consensus as to what makes a tumor cell, to which one can only respond that cancer is jolly complicated.

Making a simple statement about what basically makes a normal cell into a tumor cell was not the problem for Hanahan and Weinberg—it's mutations. We've known for some time that specific mutations or groups of mutations are often associated with some cancers. This was first established for some hereditary colon cancers, as we've mentioned, and subgroups of many types of cancer are now being defined on the basis of their complete mutational pattern revealed by whole genome sequencing (chapters 8 and 9). Nevertheless, there are no absolute rules. In other words, many mutational events have been found in cancers, and different combinations can result in cancers that are clinically indistinguishable.

The requirement to accumulate a specific hand of mutations before a cancer really gets going is, of course, the reason why most cancers don't appear until late in life. If this idea of cancers selecting groups of mutations, so to speak, and using them to their advantage sounds familiar that's because, as Hanahan and Weinberg pointed out, it closely parallels the concept of Darwinian evolution. That is, the development of a tumor cell may be compared with the evolution of a species: the driving force for both comes from the acquisition of specific mutations and the selection of those that help survival and proliferation until the cancer cell gradually acquires a growth advantage over normal cells. The cancer always requires its host to survive (it's a parasite), but eventually it is capable of diverting nutrients from the host to support its own uncontrolled expansion. This is why cancer patients may continue to eat normally but nevertheless lose weight—the observation that led Otto Warburg to call it the "wasting disease" some eighty years ago.

We've mentioned the vast array of mutations that can drive cancers, and you may have wondered why, if it's only about getting cells to grow when they shouldn't, it has such complex ways of going about it. The evolution parallel provides something of an answer. We mentioned our fishy past and the cellular trial and error that went on in the primeval ooze until some little guppy or other managed to wriggle onto land and keep breathing, the first "fish out of water." We might also add that evolution has parallels with my rare excursions into DIY to keep my house from collapsing. There's not even a short-term plan never mind anything grander. The bathroom tap's dripping: Can we unearth any old bit of rubber that'll fix it for another month or two? The first shoots of beans have emerged from the winter soil: grab some netting and a few sticks—doesn't matter what it looks like or that any fox would snigger at such a feeble barrier—the only requirement is to keep those pesky pigeons at bay for long enough to give the veggies a chance of growing into a human dinner instead of an avian breakfast. What matters in these cases—from my pigeon-free garden to some errant cell's ability to keep growing is, What works? What matters is that when a way is found that does work—when I find an effective way to keep pests out of my tomatoes or when that cell finds an efficient way to keep feeding itself—both of us are going to keep doing what works.

CELLS BEHAVING BADLY: CANCER CELL CHARACTERISTICS

The real triumph in defining what makes a tumor is to have kept one's head sufficiently above the avalanche of data that has been generated in the molecular age to piece together just what it is that these mutational events *do* to the way a cell works. From all this gene mining we now have a fairly detailed view of the molecular biology that underpins the development of tumor cells. The really important question, of course, is what that does to the way the cells actually behave. At the end of chapter 6 we listed the key changes: (1) cancer cells need much less in the way of external growth signals than do normal cells; (2) they ignore external signals telling them not to grow; (3) they can grow indefinitely; (4) they avoid suicide, a powerful anti-cancer strategy when cellular DNA is damaged; (5) they can persuade new blood vessels to grow so they have a supply of food and oxygen; (6) their metabolism changes so they don't need as much oxygen; (7) they cause inflammation and immune responses and then subvert these so that, rather than eliminating the tumor, they actually promote its growth; and (8) they can spread from their primary site to other places. Let's look at each of these characteristics in a little more detail.

1. Deaf to the World

The Greek playwright Sophocles is held to be the first to record the notion that messengers who bring bad news may be so unpopular with the recipient that they come to a sticky end. Cancer cells have a novel take on this strategy because *they* are the bad news, and one way they ensure continuation on their malevolent course is by largely ignoring messages from the outside world. Animals, including humans, are multicellular organisms—clumps of cells that communicate with each other so that each knows how to behave for the good of the cause, that is, so that the whole animal works. A critical and recurrent question for an individual cell is whether to proliferate, that is to grow and divide into two daughter cells, or whether to sit tight just doing its job. All normal cells rely on cues from the rest of the animal to make this decision. These cues come in the form of molecular messengers that are either directly presented to the target cell by adjacent cells, or that diffuse either from nearby cells or from further afield (via the bloodstream, for example). We've looked at how they activate one example of a signal pathway—MAPK (chapter 6).

A key feature of cancer cells is that they become somewhat deaf to these molecular cues—for example, by acquiring mutations in EGFR or the MAPK pathway so that they can function as if they were being told to divide when they are not. As ever in cancer, it's a little bit more complicated than that because they still require some help from growth factors to survive. What's more, as we shall see when we come to inflammation, tumors can persuade normal cells in their vicinity to give them a helping hand by, for example, making messengers that actually promote growth of the cancer. At the same time, as part of their re-wiring, tumor cells may even start to make their own growth-promoting factors. In the traditional manner of delinquents, cancer cells respond to those signals while remaining deaf to messages of restraint.

The first critical point about cancer cells, therefore, is that they change the way in which they sense the world around them and also change the way their environment sees them. By so doing they make themselves able to grow in a way that is largely independent of that world.

2. Eliminating the Negative

Given that many animal cells are continuously asking themselves "to be two or not to be two?" and that the push for proliferation comes from growth factor messengers talking to the cell, one might guess that there are also factors that interact with cells to tell them not to divide. That is, for each cell, as for the organism as a whole, there is a balance of proliferation that is a reflection of positive and negative signals. This is indeed the case because, counteracting the positive signals, there are anti-growth signals. In terms of signaling methods, the two work in broadly similar ways and cancer cells deal with them in similar fashion: by acquiring mutations in signaling pathways. The anti-growth messengers are proteins, just like the positive signals, but in cancer cells their message is no longer transmitted to the nucleus, or it becomes distorted so that it has the opposite effect, promoting growth and migration rather than stopping it. These inhibitory signals therefore restrain proliferation in normal tissue, but cancer cells frequently cease to respond to them. It's another example of cancer cell delinquency: selective hearing and twisting the message so that it says what you want to hear.

3. Everlasting Life

We saw in chapter 4 that you can remove a piece of tissue from an animal, break it up into single cells, and put them in a suitable solution, whereupon they will keep on reproducing for somewhere between 20 and 60 doublings of the cell population—and then stop. The state they have then entered is called senescence or replicative senescence: cellular old age. The cells are not dead, although they will eventually die, but they can exist for a long time in culture in this state of suspended animation: they have barely detectable metabolic activity and require almost nothing in the way of nutrients.

We also saw in chapter 4 that one reason for cells becoming senescent is that they have lost DNA from the ends of their chromosomes. Recall that our DNA is split up into 23 pairs of chromosomes, so that means there are 92 ($23 \times 2 \times 2$) free ends of double-stranded DNA within the nucleus. These are a bit of a problem for the machinery that replicates most of our DNA because it can't deal with the very end bits. The result of not being able to replicate the last bit of DNA is that every time our DNA is duplicated, which has to happen whenever a cell reproduces itself, our chromosomes get shorter. This might be serious at a very early stage if the DNA lost was important (e.g., encoded a protein). Presumably for this reason, chromosomes are "capped" at each end by repeated sequences of DNA called telomeres (the repeats are TTAGGG). Because telomeres don't code for protein, loss of these sequences doesn't do any harm to individual cells. Nevertheless, telomere loss would be incompatible with survival of the species if it happened in all cells and, for that reason, germ-line cells and some stem cells express telomerase that can achieve the trick of replicating the ends of chromosomes. In all other types of cell (that is, somatic cells), however, telomerase is almost undetectable—its gene is still present, of course, but it's almost completely "switched off," never to be turned on again. Never, that is, unless the cell becomes a tumor cell. About 90 percent of primary tumors make substantial amounts of telomerase, which means they can maintain the length of their telomeres at the ends of chromosomes, thereby escaping the finite number of doublings of normal cells. They can grow indefinitely.

We observed just now that "telomerase is almost undetectable" in somatic cells, a statement that reflects what until recently was generally held to be the case. It has emerged, however, that it *is* possible to detect some telomerase activity in most cells. This suggests that, yet again, nature works by balancing forces—in this case chromosome length. On the one hand, you might wish your chromosomes to stay the same length so that you don't grow old, but we know that very active telomerase is a major contributor to cancer development because it helps to "immortalize" cells.

As with much in nature, you might assume that, given this new information about the seeming capricious nature of telomeres, your precise position on the life and death high wire is out of your hands—you're in a game of genetic roulette. Amazingly, that may not be quite true as it is emerging that we may be able to adjust our telomerase activity, and hence the rate at which our chromosomes disappear, by changing our lifestyle. In other words, telomerase may turn out to be another biochemical marker for stress. Individuals who endure stress over prolonged periods (e.g., people who care for patients with severe disabilities or who are HIV positive but without overt symptoms) have lower telomerase activity and shorter telomeres than their less stressed brethren. This apparent capacity to tweak our telomerase even extends to identical twins: if one has a sedentary lifestyle (he's bone idle) and the other is active ("works out"), this can be reflected in the relative activity of their telomerases (no prizes for working out which way round).

4. Dodging Death

We have seen that normal cells carry on a balancing act with regard to proliferation, and that this is a feature of all multicellular organisms. There's nothing particularly startling about this concept, although the scale and range of activity involved is staggering. Just to stay as we are, human beings make 2 million new cells every second. I sometimes point out to second-year biochemists that if they totted up all the cells they made while sitting in their lectures, they'd have just about enough to make a new student, which at least makes them realize how comprehensive the course is. The new student would be a distinctly odd blob, however, and seriously short on brain power as, after humans complete development, we make almost no new neurons, skeletal muscle cells, or heart muscle cells. The inability of heart cells to proliferate is one reason why heart attacks are so serious—any resultant cell loss cannot be replaced. However, epithelial cells—the ones that form the lining of the structures of our body, including the skin, lung, and intestine—proliferate rapidly throughout life. Our blood vessels have a specialized lining called the endothelium made of (endothelial) cells that usually just sit there, only switching on to make more of themselves if we get injured, as part of the process that repairs our blood vessels. This turns out to be very important in the development of solid tumors, and we'll come back to it shortly.

So some cells are dividing like mad, others have completely lost the capacity to divide at all while some, like the lymphocytes that make antibodies, are just waiting for the right cue to switch on division. None of that is too surprising. What is, perhaps, unexpected is that our cells have an inbuilt system for committing suicide that we spoke of earlier. They have a capacity for self-elimination (e.g., after DNA damage) called apoptosis and it has emerged over the last few years that this suicide program is inhibited in most, perhaps all, cancer cells.

There are two major ways in which individual cells can be killed: necrosis or apoptosis. In necrosis (from the Greek for "dead": *nekros*) some parts of the cell are broken down but, in essence, it simply bursts and releases its contents to the surrounding tissues. The chemicals released by the cell may then cause inflammation at the site. Necrosis can be caused by infection or injury. It occurs, for example, in a heart attack when the blood supply to the heart is interrupted. For the same underlying reason it can also be important in cancer: that is, the central regions of solid tumors may have a poor blood supply—we'll speak more of this shortly—and when this happens, tumor cells can be killed off because of the lack of blood-borne nutrients. When necrosis blocks the blood supply to the limbs, it can lead to gangrene, a condition in which the body tissues themselves start to decay: the affected region turns black because of the iron released from hemoglobin after red blood cells have been broken down. These days gangrene is rare in people who are generally fit, although it can occur in the feet of individuals with diabetes or with arteriosclerosis, again because of inadequate blood supply.

Apoptosis was first given a biological meaning by John Kerr, Andrew Wyllie, and Alastair Currie at the University of Aberdeen in 1972. They observed that cells sometimes break up into fragments and that these are then either shed and carried away by the circulation or taken up by other cells and rapidly degraded. In other words, the evidence of cell death is rapidly removed (in contrast to necrosis), which was why this event had been so hard to pin down. Kerr, Wyllie, and Currie showed that apoptosis is involved in cell turnover in many healthy adult tissues and also in the elimination of cells during development. On first acquaintance, the idea that there is a suicidal component built in to the normal growth of animals might seem counterintuitive. However, many parts of our body attain their final form by shedding bits. Nowadays many parents-to-be will know this if they have ultrasound scans of their unborn baby, which reveal how our fingers develop from what is really a web until the destruction of cells removes the tissue between the digits. A controlled program of cell elimination is therefore part of normal development.

More relevantly from the cancer point of view, Kerr and his colleagues found that apoptosis could always be detected in untreated malignant tumors. This was consistent with earlier studies showing that not only were cells lost in growing tumors but sometimes they were dying almost as fast as new tumor cells were being made. It might come as an even bigger surprise to discover cell elimination in tumors than to find it going on during normal development. However, this may be a useful point at which to remind ourselves that normal tissues are in a balanced state between proliferation and cell loss, and that tumors merely represent a slight perturbation of normality.

Apoptosis is therefore a regulated response to external signals that makes the cell commit suicide. Because it is a "programmed" response, the cell digests much of itself, including its DNA, rather than simply exploding, which distinguishes apoptosis from necrosis.

The fact that we lose cells at a rapid rate shouldn't really be a great revelation. You can easily show, preferably in the privacy of your own home, using no more complex a piece of equipment than a finger nail to scrape your own skin, that human beings shed a lot of skin all the time—actually somewhere in the region of 30,000 cells every minute. This loss is from our outer layer (the epidermis) and by the time cells reach it, they are mostly dead. Because we don't disappear, we must therefore make skin cells at the same rate as we lose them, so the whole business of keeping ourselves clad in skin is an equilibrium between cell loss and production. In fact the half-life of skin cells is about thirty-five days; that means that half of your skin today will be in the Hoover in a month's time. So this is big business on the cellular front, and it's not surprising that big business has cashed in: there's a huge number of commercial concoctions on the market that purport to enhance either the rate of skin cell renewal or exfoliation, all with the aim of persuading us to part with money in the pursuit of beauty. Quite how effective they are is a debatable point. Skin loss isn't apoptosis, but it is another example of balance or what biologists call homeostasis, in this case in cell turnover. Perhaps the other most familiar example of homeostasis is red blood cells. These have a half-life of 120 days: rather than being shed, which would make even more of a mess of the bedclothes than does skin, time-expired red cells are broken down in the circulation and consumed by scavenger cells (macrophages). The daily cell loss is 10^{11}—maintaining the equilibrium requires us to make 200 billion blood cells per day—which is why we can cheerfully contribute a regular pint to the blood bank.

We now have a fairly clear picture of how apoptosis is regulated at a molecular level and, in particular, how the activation of the P53 tumor suppressor protein in response to DNA damage can switch on the destruction of mitochondria, which leads to the dissolution of the entire cell.

Dormant Tumors

Thinking about cell proliferation versus cell death leads to the idea that minitumors might actually form and then stop for whatever reason, remaining static as "dormant tumors." In fact, such microscopic growths were first detected early in the twentieth century and, more recently, autopsies of road traffic accident victims came up with the rather disturbing finding that many adults have a significant number of static cancers (also known as *in situ* tumors). These contain about 100,000 cells, occur in a variety of organs and tissues, and would normally be undetectable—it was just that accidental deaths had provided tissues for pathological analysis. These microtumors were clearly dormant: their carriers died in accidents and had shown no signs of cancer. Knowing what we do about the time course of cancer development, we can be sure that most of them would not have gone on to produce cancer for many more years or even decades. Clearly, the dormant tumors had spontaneously ceased to grow, and the presumption is that they lacked our next cancer cell feature: the capacity to provide themselves with a blood supply.

5. Bleeding the System

We have already encountered the mighty J. B. S. Haldane, who was described by Peter Medawar as "the cleverest man I ever met." Yet another of his claims to fame takes us to the heart of why new blood vessels are essential for tumors to grow. This observation, made in his essay "On Being the Right Size," has since been christened Haldane's principle. It states that sheer size very often defines what bodily equipment an animal must have: "Insects, being so small, do not have oxygen-carrying bloodstreams. What little oxygen their cells require can be absorbed by simple diffusion of air through their bodies. But being larger means an animal must take on complicated oxygen pumping and distributing systems to reach all the cells." Haldane, in addition to founding the science of population genetics, knew a fair bit about insects, including that there are 400,000 species of beetle compared with only 10,000 species of mammals—knowledge he employed in his devastating response to the question of what his studies had led him to conclude about the Creator: "If he exists, God has an inordinate fondness for beetles."

Nowadays, most people have grasped the idea that all animals need food and oxygen to survive because the individual cells they're made of have the same requirements. There is a difference, of course, in that your dinner reaches most cells in a somewhat processed form, broken down into simple components (sugars, amino acids, fatty acids, etc.) that by then are often referred to as nutrients. These, along with oxygen, are delivered in mammals by the bloodstream. For oxygen, more precisely, it's red blood cells that do the carrying: they outnumber all other blood cells by about 1,000 to 1 and they're red because they're essentially sacs of the iron-containing protein hemoglobin, which carries oxygen. Cells require oxygen mainly to enable them to make enough ATP (adenosine triphosphate) to drive the biochemical reactions that keep them going. Oxygen diffuses from red cells in the bloodstream; to get enough of it, other cells need to be within about ten cell widths of a blood vessel.

In this respect, cancer cells are just like their normal brethren. By definition, however, they are abnormal: they're a pathological growth that the body isn't set up to accommodate and won't provide with a blood supply in the way that happens to a tissue during normal development. This is a major defense against cancer: even though a "microtumor"—a small cluster of cancer cells—may have established itself, it cannot grow much beyond about one millimeter in diameter unless it can acquire a blood supply. Only relatively recently have we discovered that cancer cells can actually make their own blood supply. In essence, what happens is that something causes tumor cells to start releasing a signal (another hormone). This turns on enzymes that can chew through the walls of blood vessels. The damage caused in turn acts as a signal, switching on growth of the cells that line the vessels (endothelial cells). As these multiply, they migrate through the gaps in the vessel wall and lay down new branches—

so the whole process is often called "sprouting" or, more formally, "angiogenesis." Sprouting blood vessels then infiltrate the microtumor, providing the oxygen and nutrients to kick it off.

In switching on angiogenesis, tumor cells are not doing anything novel as far as the organism is concerned. This process, in which new capillaries sprout from pre-existing vessels, is crucial for embryonic development but is almost absent in adult tissues, other than when repairs are needed, for example, in wound healing or during the female reproductive cycle. Much less usefully, angiogenesis occurs as a pathological condition in over seventy diseases, including heart and vascular disease, rheumatoid arthritis, Crohn's disease, psoriasis, endometriosis, and proliferative retinopathy (a common consequence of diabetes). The critical feature of angiogenesis is that signals activate the proliferation of the endothelial cells that line the vessels of our circulatory system. Because this lining is just one cell thick, it is quite fragile and, therefore, as we shall see, it can easily be damaged, thus providing a route from the outside into the circulation and in the opposite direction. Lining the vascular system is the only job that endothelial cells do: even though the lining is only one cell deep, such is the extent of our circulatory system that if you could arrange all our endothelium as a carpet, it would cover about one quarter of a football field.

The strategy of persuading normal blood vessels to behave abnormally by expanding to service a growing tumor might be described as "cunning" were it not for the implication of prescience—the suggestion that tumor cells "know what they're doing" and plot the subversion of their host accordingly. They do no such thing, of course, and the astonishing feat of instructing the architecture of the body to suit its own needs only reflects the limitless opportunities generated by evolutionary trial and error.

Feeding Tumors

Tumor cells, if they are to grow significantly, need their own vascular system. This is thought to be a critical step in tumor development that must happen first in the primary tumor and then subsequently when cells from the primary have spread to secondary sites, if those metastases are to expand. Of course the formation of a vascular network within a tumor also creates a transport system waiting, so to speak, for metastatic cells to make their escape.

Regulating Blood Vessels

Like so many other facets of living systems, growing new blood vessels is a delicate balance: too much or not enough at any moment might be fatal. You might expect it, therefore, to be a multistep process and to be under continuous regulation by both positive (pro-angiogenic) and negative (anti-angiogenic) factors. The hormone that we met just now turning on sprouting is a protein called VEGF (vascular endothelial growth factor). It's the most potent angiogenic factor known and, when it's present in the neighborhood of tumors, it does what its name suggests: makes the endothelium grow. VEGF binds to specific receptors to "turn on" the cells (another example of the chapter 4 story of growth-factor signaling). The growth of tumor vessels is an obvious therapy target, and we'll return to that in the last chapter, but it's also been used to predict outcome: high levels of VEGF give a dense network of tumor blood vessels, and that is associated with poor prognosis in a variety of human tumor types.

The first anti-angiogenic protein to be isolated is called angiostatin because it stops the growth of endothelial cells and can prevent human cancer cells from forming tumors when they are inoculated into mice. Over twenty angiogenesis inhibitors have now been discovered and it's notable that some of their genes live on chromosome 21—the one present as three copies instead of the normal two in people with Down's syndrome. Strikingly, Down's syndrome individuals have about half the normal lifetime risk of developing most cancers (although such children are between twenty and thirty times more likely to get leukemia), which may be due to the result of their having higher levels of anti-angiogenic factors.

6. Eating for Two

Tumor Blood Flow and the Flexible Cancer Cell

All the talk about blood simply shows that for tumors to grow beyond a relatively small size, be they primary or secondary, they need to generate their personal supply, albeit via vessels made from normal cells. However, tumors *are* abnormal growths and it's not surprising that the blood vessels they create are also somewhat weird. That is, usually they don't have any discernable pattern: sometimes they just stop like a kind of cul-de-sac, sometimes blood flows into them from both ends, generating a form of traffic chaos and, in general, they are leaky and tortuous compared with normal tissues.

This picture of chaotic structure and flow suggests that some parts of a tumor may get less oxygen than others; not only is this true, but cells within tumors often survive on much less oxygen than normal cells in adjacent tissue. However, the center of a growing tumor can become so hypoxic—that is, oxygen-deprived—that the cells die, forming a necrotic core, while the outer regions of the tumor continue to grow. This has led cancer cells to evolve some quite remarkable adaptations to help their survival. These revolve around the metabolic pathways that convert glucose into ATP, the principal energy currency of the cell.

Most of us are aware of the importance of carbohydrates (molecules made up of carbon, oxygen, and hydrogen) in our diet. The critical sugar in carbohydrates is glucose, and most organisms (more or less everything from bacteria to humans) use it as a fuel. It's a fuel for cells because it is broken down (oxidized) by a series of chemical reactions to make ATP. ATP is important because it can supply the chemical energy to drive many reactions in cells. There are a lot of steps (distinct chemical reactions) required for this, but the important point is that these can be split into two main stages.

In the first, glucose is broken down into two molecules of pyruvate: this pathway (glycolysis) is an anaerobic reaction (meaning it doesn't need oxygen; it parallels what happens when yeast makes alcohol). The final step in the pathway is catalyzed by the enzyme pyruvate kinase—in yeast, two further reactions convert pyruvate to ethanol: in us, the enzyme lactate dehydrogenase turns it into lactic acid). The most important thing, however, is that glycolysis produces two molecules of ATP for every glucose that is turned into pyruvate.

The second stage in the destruction of glucose—it's actually being turned into water and carbon dioxide—makes much more ATP per glucose molecule consumed. It also differs from the first stage in that it takes place inside a specialized organelle—the mitochondrion (chapter 3)—and it uses oxygen. That makes the second stage "aerobic" (it's oxidative phosphorylation), and it gives thirty ATPs from every glucose molecule.

Bearing in mind that cancer cells are growing and multiplying, and therefore using much greater amounts of ATP, one might suppose that they would use the highly efficient aerobic pathway whenever possible, but they don't.

The first person to spot that there was something odd about metabolism in cancer cells was the biochemist Otto Warburg, who eventually demonstrated that cancer cells get most of their energy from glucose using the first stage, rather than the second that uses oxygen, and they do this even when there is plenty of oxygen available. Presumably, this reflects the fact that tumor cells have adapted to being fed by the disorganized blood supply they create, which gives rise to regions of low oxygen pressure. Warburg suggested, incorrectly, that cancers actually occur because mitochondria develop faults—in other words that it is a disturbance of metabolism that drives tumor formation—whereas we now know that metabolic perturbation is just one of the consequences. However it's come about, it represents a startling adaptation by tumor cells. Nevertheless, Warburg's discovery, reported in the 1920s, languished largely ignored in the literature for some seventy years until technical advances began to reveal the extraordinary behavior that lies behind his observation.

Warburg may have been somewhat miffed by this, although he was doubtless consoled by his Nobel Prize, but even without much heed being paid to his findings, some enlightening observations were made in the intervening period. We've seen how damaging radiation can be to cells, but in the 1940s and 1950s it was shown that the lethal effect of radiation used to treat cancers can largely be blocked if the tissue has previously been exposed to low levels of oxygen. This provocative finding was joined by others showing that a variety of cell responses, including laying down blood vessels, could be affected by oxygen levels, all of which focused on a critical question.

How Do Cells Sense Oxygen Levels?

The glimmerings of a molecular explanation began to emerge when several genes were shown to be switched on to make RNA and protein after cells were exposed to low levels of oxygen (one of these was VEGF). Recall that genes are regulated by proteins (transcription factors) that bind to the regulatory regions of their DNA sequence. It was therefore not a complete surprise when, in the 1990s, a number of transcription factors were discovered that are expressed in response to hypoxia (called, of course, hypoxia-inducible factors, HIFs). These were the regulators of the oxygen-responsive genes and, indeed, it quickly emerged that the regulatory regions of those responsive genes had specific sequence motifs to which HIFs could bind—and hence control gene expression.

That, of course, provided a means by which reduced oxygen levels could turn on genes. What it didn't do was explain how oxygen is "sensed" to affect the levels of HIF proteins. To do this, there has to be a protein (or proteins) that actually react with oxygen. HIF proteins don't do this, but it has transpired that their very existence depends on proteins that do (they're enzymes). When there's a reaction between two chemicals, the speed of the reaction depends on their concentrations. If one of them is oxygen, the less there is (the lower the oxygen level), the slower the reaction—and hence the lower the amount of product(s). If an oxygen-using enzyme makes something that breaks down HIF proteins to inactivate them, you have an oxygen-sensing system that can regulate the levels of HIFs. You can guess what enzymes that use oxygen directly are called (oxygenases), and some of them do indeed control the amount of HIF proteins in cells.

In fact, oxygenases are not as mysterious as you might suppose: you do an experiment with one every time you take an aspirin—and it's a fair bet that you do occasionally, given that the world seemingly eats 40,000 tons of the stuff each year. Aspirin is a non-steroidal anti-inflammatory drug (NSAID): it works by stopping a reaction needed for cells to make messengers that control several important responses including blood clotting (which is why aspirin is sometimes given to prevent heart attacks and strokes), inflammation, and the sensitization of pain receptors (which is why you took your aspirin in the first place). The British pharmacologist John Vane won a Nobel Prize and acquired a knighthood for showing that aspirin works by blocking an oxygenase. But how do oxygenases control HIFs? The answer is in a rather roundabout way: When normal levels of oxygen are present, HIF proteins are made but are rapidly tagged; this label is recognized by a sort of cellular garbage collector that gets rid of proteins, so that HIF proteins normally have a very short lifespan. The enzyme that does the tagging is an oxygenase and, by using oxygen, it acts as a concentration detector in the cell. As oxygen levels drop, so too does the activity of the oxygenase; HIFs are no longer targeted for destruction, their levels rise, and their target genes are switched on. In short, HIF proteins have the job of switching on genes that help cells make new blood vessels (e.g., VEGF) but HIFs themselves are broken down rapidly as oxygen levels rise—they're indirect oxygen sensors but their action enables tumors to thrive in hypoxic locations.

The Return of Otto Warburg

We've seen that solid tumors are a bit of a mess: the disorganized blood supply means that some of the cells have plenty of oxygen whereas others are starved. Because cells can sense this difference and express different genes—make different proteins—in response to the varying oxygen level, tumors are a real mixture. The Warburg effect, discovered all those years ago, reflects the fact that some tumor cells use glycolysis as their major source of energy, meaning that they have adapted their metabolism to their circumstances. In doing this the cells make a lot of lactic acid and release it into their surroundings. It's exactly what happens when we exercise our muscles and get a cramp—that's caused by the acidic effect of too much lactate being released into the bloodstream.

But we saw in chapter 4 that, for the most part, a chemical can only get in or out of a cell through its own private doorway—that is, there has to be a protein carrier tailor-made for each chemical. Tumor cells making a lot of lactate must therefore provide the means for its exit—which they do by making a lactate carrier that spans the membrane. Lactate, in effect, is a chemical key that unlocks its own exit from the cell. Almost incredibly, other cells in the confused environment of the tumor that find themselves in oxygen-rich regions make a slightly different carrier: this also carries lactate but in the opposite direction, allowing it into those cells to be used as the fuel for oxidative phosphorylation, the most efficient way of making ATP. This extraordinary adaptability enables tumor cells to adjust their metabolism to their local surroundings in the constantly fluctuating environment of a growing tumor.

Even the phlegmatic Warburg would have been impressed by the amazing cellular gymnastics that enable tumors to make the most of their uncertain situation—and perhaps he would be quietly pleased that we now recognize the more prosaic description of "metabolic perturbation" to be a characteristic of most, if not all, cancers.

7. Inflaming the Situation

We're all aware that when something abnormal happens to our bodies, be it infection by germs or damage to tissues from whatever cause, the infected area almost always becomes inflamed. Given that tumors are distinctly abnormal, albeit arising by corruption of the normal, a rather obvious question would be "does inflammation have anything to do with cancer?" The answer is "yes" to the extent that inflammation is involved in most, perhaps all, cancers. Over twenty-five years ago Harold Dvorak described tumors as "wounds that do not heal," but it is only relatively recently that the importance of inflammation in cancer development has begun to be fully appreciated. Inflammation is really the first sign that the immune system is being activated. The symptoms that we've all experienced when we cut our finger or bump a shin against a table leg arise because cells that are infected, injured, or stressed (and activated "cancer genes" are a stress) release chemicals that signal their plight. Tissue damage also makes blood vessels leaky, and the result of all this is that white blood cells of the immune system home in on the area. The point of all these responses, of course, is to destroy the cause and repair damaged tissue.

Two quite well-known conditions that reflect the link between sustained infection and inflammation are Crohn's disease (not to be confused with irritable bowel syndrome, which isn't associated with inflammation) and inflammatory bowel disease, which affects the lining of the intestines and includes ulcerative colitis. These are quite common (in the US about 1 in 1,000 have Crohn's disease and 1 in 500 ulcerative colitis), and they carry an increased risk of bowel cancer. They are autoimmune diseases—that is, they arise because lymphocytes make antibodies that recognize proteins on our own cells ("self" recognition). This means that our immune system has lost an element of control and attacks cells that it should leave alone.

We commented in chapter 3 on the fact that chronic infection can promote the development of cancers and indeed some 20 percent of cancers worldwide are caused by bacteria, viruses, and parasites. This has led to various oncogenic viruses and infectious organisms such as *Helicobacter pylori* being classified by WHO as carcinogens. DNA tumor viruses can inactivate tumor suppressors, and the inflammation caused by *H. pylori* can make the host produce chemicals that can mutate DNA. These all have in common the fact that sustained infection invariably causes inflammation, and that can lead to cancer.

The Tumor Neighborhood

The cells of animal tissues are supported by a framework of fibers within which live various other types of cells. Because of the inflammatory and immune responses activated, the vicinity of solid tumors attracts a wide variety of cell types, most notably lymphocytes and macrophages. These cells were first seen in tumors in the nineteenth century by Rudolf Virchow, the German "father of modern pathology"; over fifty years ago, it was suggested that malignant cells are generally killed by the immune system, which is why so few develop as metastases. More generally the notion is that "immunosurveillance"—the action of the immune system—kills cancer cells and is hence an important protective mechanism. Under some circumstances, rather than completely eliminating a cancer, the immune system may merely suppress proliferation, thus being another contributor to the dormant tumors that we discussed earlier.

It's only recently that we've been able to pick out individual types of cells and begin to unravel the full complexity of the cellular mosaic comprising the tumor microenvironment. The cellular groupies recruited by the tumor are a cosmopolitan lot, and they infiltrate and closely interact with the cells of the tumor. It's a two-way communication, both tumor cells and normal cells

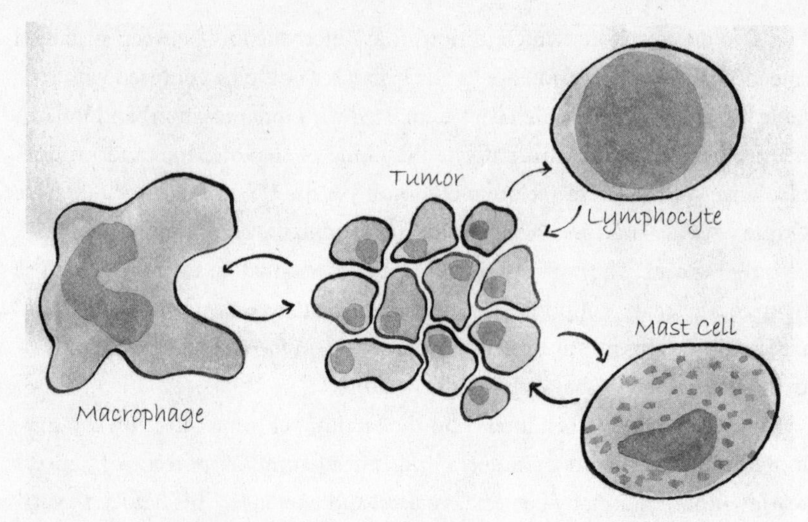

The tumor neighborhood. *Two-way communication between host cells and tumor cells. Various types of white cells (lymphocytes, mast cells, macrophages) are attracted to the tumors as part of the inflammatory and immune responses. These recruited cells may eliminate tumors. However, tumor cells that manage to evade this defense acquire the capacity to signal to the recruited cells, causing them to switch their action so that they now actively support tumor growth.*

releasing the messenger proteins involved, and there is much evidence that this response can inhibit cancer and eliminate malignant cells. Interferons and interleukins are major types of messenger, and two members of these families that stimulate the immune system are now Food and Drug Administration (FDA)-approved treatments for a variety of types of cancer, particularly leukemias and lymphomas. The other side of the coin is that patients taking immunosuppressants (to treat autoimmune diseases or after transplants) have an increased risk of some forms of cancer.

However, despite the immunoprotection that can be conferred, the fact that tumors *do* develop indicates that this can be overcome by determined cancer cells. Indeed it's now clear that the tumor microenvironment can change sides, so to speak, and become immunosuppressive and tumor-promoting. This switch should not be altogether surprising because, although the first role of the inflammatory response is to remove the damaging agent, a second feature is to restore the tissue to its normal state. This, of course, requires new cells—and this means that the initial anti-tumor response harbors the capacity to switch on survival and growth signals, the very drivers of tumor progression.

The overall effect of signaling in the tumor environment is therefore a reflection of forces that are themselves unstable. That is, whether specific cells and messengers are tumor-promoting or tumor-suppressive depends on the context—for example, the stage of tumor development or the level of hypoxia.

So it isn't that tumors are entirely deaf to the outside world, rather that they are selective in their hearing and subversive in their recruitment of host cells that become involved in two-way dialog and are subsequently converted from enemies of the tumor to rabid supporters.

8. Happy Landings

The seven features of tumors we've considered are characteristic of cancer cells but may also occur to some extent in benign tumors. The size to which benign tumors can grow indicates that they acquire a degree of independence from growth factors, both positive and negative, don't commit suicide, continue to divide, have a degree of metabolic abnormality, and keep themselves supplied with blood. Moreover, inflammation is also a common cause of benign tumors. Therefore, although these seven properties are generally associated with tumor development, they don't absolutely distinguish between benign and malignant growths. In short, the distinctions we've drawn thus far between malignant and benign, although very informative, have been to some extent a matter of degree. The eighth property, however, absolutely defines malignant cancers: the capacity of cells to invade surrounding tissue and spread to other sites.

Primary and Malignant Tumors

Primary tumors generally have quite sharply defined edges and, much as with benign tumors, they tend to be a problem only if their size affects the working of nearby organs or if they make things that affect normal physiology (e.g., hormones such as insulin). They are usually treatable by surgery, which, for most early-stage tumors, is highly effective. When tumors recur at the site of the primary, it is generally because not all the tumor cells were removed. This means that, in the main, primary tumors do not kill: most cancer deaths result from metastasis, which, as we have already discussed, is what happens when cells acquire the capacity to invade surrounding tissue and spread to other sites in the body—that is, to become malignant.

Metastasis is poorly understood, but the steps that a cell has to take to become metastatic are simple: (1) migrate from the primary tumor, (2) burrow

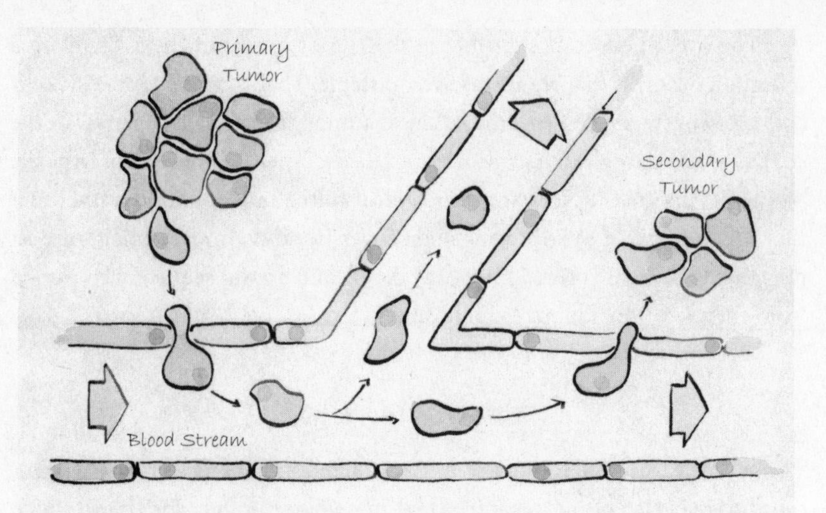

Cells from a primary tumor entering the circulation, escaping at another site and forming a secondary tumor (metastasis).

through the wall of a blood or lymphatic vessel, (3) survive in the circulation until it sticks to a target site, (4) burrow through the wall of the vessel, this time to escape, and (5) start to grow in its new location. There's no need for the steps to be a continuous sequence—indeed, we've noted that some tumor cells lie dormant in secondary sites for many years without their growth being reactivated. As ever with cancer, there are variations on the basic theme. For example, ovarian cancer usually spreads by cells shedding into the abdominal cavity and then sticking to the surface of the liver, and some brain tumors tend to infiltrate normal brain tissue rather than spread via the circulation. Certain grades of sarcomas, rare malignancies that develop in the bones or soft tissues, do not generally metastasize.

To take the first steps down the metastatic road, a cell must start to release enzymes that break down the fibrous network surrounding the tissue in which they were born. It can then enter the circulation and trundle round until it finds a bit of the vessel through which it can escape into normal tissue. Almost inevitably, many people with a primary tumor only see a doctor when it starts to cause obvious symptoms (pain or bleeding, for example)—which usually means it's already metastasized. Physicians are then usually presented with the problem of how to deal with secondary tumors for which, at the moment, they have very little in the way of weapons. While not directly bearing on that problem, in getting a handle on metastasis it would be a major advance if we could find out how tumor cells select their target sites. Although we know that some types of cancer cell do home in on particular tissues (e.g., for breast tumors a common sequence is lymph nodes followed by bone, liver, and lung), for the most part we have little idea how wandering tumor cells select their target sites.

Early Ideas about Metastasis

Metastasis is easy to define—it's the spread of cancer cells from their primary site to other locations—and it's been almost two hundred years since biologists recognized that cancers might spread in this way. Nevertheless, even at the beginning of the twenty-first century, the honest answer to the question "What controls the spread of tumors?" is "We don't know." There are two reasons for this sustained state of bafflement. One is that now we at least know enough to realize that metastasis is mighty complicated (i.e., we *are* baffled). The second is that, as ever in science, progress has been limited by the available technology and it's only recently, for example, that we've been able to identify specific proteins in animal models of tumor progression.

We mentioned in chapter 1 that René-Théophile-Hyacinthe Laënnec was the first person to record from his observations on melanoma that cancer could spread—metastasize—to secondary sites. The first use of the term "metastasis" is credited to a French surgeon, J. C. A. Récamier, who succeeded Laënnec as a member of the Collège de France. To be more precise, being French, he used the word "*métastase*" to describe the formation of secondary tumor growths that had derived from a primary breast tumor. The first person to show that metastasis occurs because cells leave the primary site and spread through the body, however, was the German surgeon Karl Thiersch. This was at odds with the notion that had been proposed by Rudolf Virchow that tumors released a "juice" that could somehow change normal cells at the secondary site into metastases. As Virchow was very famous (he discovered leukemia and is generally considered to have established the subject of pathology) and Thiersch was not, it was only to be expected that it would be a while before anyone paid much attention to Thiersch's idea, notwithstanding the evidence of Laënnec and others. Thiersch eventually triumphed although, coming right up to date, the debate has resolved itself as a typical scientific argument in that everyone was right and everyone was wrong. Though notable, this wasn't Thiersch's only claim to fame. His experience as a surgeon in various wars of the second half of the nineteenth century had led him to an improvement in the sterilization method of Joseph Lister by replacing carbolic acid with salicylic acid. What's more, he became the son-in-law of the aforementioned Justus von Liebig.

Once you accept the notion that cells migrate from a primary tumor to form a secondary, the obvious question is: "How do they know where to stick?" Or, as Stephen Paget phrased it more elegantly in a landmark paper of 1889: "What is it that decides what organs shall suffer in a case of disseminated cancer?"

The simplest answer would be that it just depends on anatomy: that is, cells leave a tumor and then stick at the first tissue they meet in the way that can happen in breast cancer. However, Paget had spotted that quite often this simply didn't happen and in his paper he described his own evidence and summarized the work of a number of other luminaries to show that "the distribution of secondary growths was not a matter of chance." Paget's specialty was breast cancer and in 735 fatal cases he had found 241 with liver secondaries, 70 lung metastases, 30 kidney metastases, and 17 metastases to the spleen, with bones also being a frequent target. Paget gives full credit to the contributions of others and particularly records Ernst Fuchs's uncommonly prescient suggestion in 1882 that certain organs may be "predisposed" for secondary cancer. All of which led Paget to a botanical analogy for tumor metastasis: "When a plant goes to seed, its seeds are carried in all directions; but they can only live and grow if they fall on congenial soil." From this, then, emerged the "seed and soil" theory as at least a step to explaining metastasis. The great strength of the seed and soil view is that it conveys interplay between tumor cells and normal cells, their actions collectively determining the outcome. As we shall see shortly, that is at the center of our current picture of metastasis. Rather charmingly, Paget concluded his 1889 paper with: "The best work in the pathology of cancer is now done by those who are studying the nature of the seed. They are like scientific botanists; and he who turns over the records of cases of cancer is only a ploughman, but his observation of the properties of the soil may also be useful."

Paget's self-effacing nature may have owed a good deal to his coming from one of those super-talented families that perhaps discouraged its members from becoming uppity. His Uncle George had been Regius Professor of Physic at Cambridge and his dad, Sir John, is immortalized, at least for physicians, by having identified a number of diseases that bear his name. These include Paget's disease of the breast, a malignant condition of the nipple that is almost always associated with underlying breast cancer, although it is quite rare (fewer than 5 percent of breast cancers).

How Do Tumor Cells Become Metastatic?

If we accept that the sites of metastases are not simply a consequence of anatomy, perhaps we should take a step backward and ask how cells become metastatic. Unfortunately, that too remains a seriously murky area, but it seems probable that a subset of the earliest mutations acquired by cancer cells—which give them a proliferative advantage—are also able to promote metastasis in later generations of tumor cells after they've picked up further mutations. This idea is intuitively attractive because it implies that the primary tumor, or at least some of its cells, gradually develops the capacity to make the critical step of dispatching cells to other locations that will, in the end, be fatal for the host. However logical that model appears, as ever there are observations indicating that at least some cancers do it differently.

Occasionally, metastatic growths appear when no primary tumor can be detected. These are classified as "cancers of unknown primary" and they are not that uncommon, falling in the top-ten chart of diagnoses. Whatever the molecular events that promote these tumors are, they clearly do not require prolonged development of a primary. Regardless of when, in the development of a primary tumor, cells become able to leave and form secondaries, we can now look at all the genes being expressed in cells. From this it's clear that the patterns differ between metastatic cells and their non-metastatic counterparts in the primary tumor. What appears to happen is that, in response to signals they receive from nearby cells or other environmental triggers, cells become less sticky, so that they detach from the primary tumor and start making proteins that chew through their surroundings, including the walls of blood vessels. Given the huge range of genetic disruption that occurs in cancer cells, there seems no reason why this shouldn't happen very early in the life of some primaries, although one might suppose that the older the tumor, the greater the chance that it will come up with a recipe for spreading.

The Morphing of Tumor Cells

With all these upheavals—different proteins being made and cells acquiring the capacity to move and chew through membranes—you might suppose that cells would change their appearance as they become metastatic. You would be quite right; it's mainly apparent in a sort of scaffold of proteins inside the cells (called the cytoskeleton) that gives them their shape and enables them to move. It can be tracked in cells grown in the lab, once again using fluorescently labeled antibodies, by the appearance of some proteins and the loss of others as the cells elongate and start to look spindle-shaped.

A very similar thing happens to stem cells in the embryo during early development—these are the cells that can grow into any type of cell and that in adults occur in the bone marrow. Quite remarkably, if you take normal cells and make them capable of forming tumors, they turn into something that closely resembles a stem cell. Because of their high capacity for self-renewal, stem cells are considered to be prime targets for the mutational assaults that produce tumors. As stem cells become progressively more genetically unstable, they may promote changes in adjacent normal cells that, in turn, develop into part of the expanding tumor.

How Do Metastatic Tumor Cells Know Where to Go?

For well over a century, Paget's aphorism of "seed and soil" pretty well summed up our knowledge of how metastasis worked, and we've had to wait until the twenty-first century for any further significant insight into the process. Perhaps the most substantial breakthrough has come from mice and the use of antibodies that we've already seen are essential for identifying specific cells by the proteins they make (despite having the same number of genes, *we* use the antibodies, not the mice). The extraordinary finding is that cells in the primary tumor release proteins into the circulation and these, in effect, tag what will become landing points for wandering cells. Extraordinary because it says that these sites are determined *before* any tumor cells actually set foot outside the confines of the primary tumor. It's a bit more complicated, however, than simply marking the target by a single type of protein. In addition to sending out a target marker, the tumor also releases proteins that go to the bone marrow. This is the place where the cells that circulate in our bodies (red cells, white cells, etc.) are made from stem cells. The arrival of signals from the tumor causes some cells to be released into the circulation; these carry two protein markers on their surface: one binds to the premarked landing site, the other to tumor cells once they appear in the circulation. It's a double-tagging process: the first messenger makes a sticky patch for bone marrow cells released by another messenger, and they become the tumor cell target. It's more of the molecular Velcro that we've already met in signaling pathways. David Lyden and his colleagues call it "cellular bookmarking."

This seems an extraordinarily elaborate mechanism for directing tumor cells to a target. Given that tumor cells cannot evolve in the sense of getting better at being metastatic—they just have to go with what they've got—how on earth might it have come about? We don't know, but the most likely explanation is that they are taking advantage of natural defense mechanisms. We've touched several times already on the point that tumor cells are, in a sense, "foreign." That is, they are an abnormal growth and therefore activate inflammatory and immune responses, just as happens if we cut ourselves and bugs get into the wound. Perhaps what is happening in these mouse models is that the proteins released by the tumor cells are just a by-product of the genetic disruption in cancer cells. Nevertheless, they may signal "damage somewhere in the body" to the cells of the bone marrow. That at least would explain why the bone marrow decides to release cells that are, in effect, a response to the tumor. The second question is trickier: Why should tumors release proteins that mark specific sites? Well, we knew before these mouse experiments that different types of tumor do indeed head for different places. The mouse models, encouragingly, showed the same effect: change the tumor type and you get a new target. In fact, this point can be made even more strongly by growing the tumor cells *in vitro* and injecting into mice just the medium in which the cells have grown—no cells, just medium containing, of course, any proteins that have been released by the cells. The result is that the metastatic sites are generated just as they would have been if a primary tumor had been sending out the protein messengers. To show that these sites still work as sticky landing strips, just inject tumor cells into the blood stream: they do indeed stick to the same sites that are targeted by cells from a primary tumor. So different messengers from different types of tumor stick (usually) to different, but tumor-specific places around the circulation.

Metastatic Footnote

Metastasizing cells are like an iceberg that breaks away from a large mass and is carried off by the ocean currents. Sooner or later the iceberg will melt and vanish; much the same thing happens to almost all metastatic cells: they get picked up and destroyed by scavenger cells in the circulation. However, the problem is that some metastasizing cells are not eliminated: they stick to their target site somewhere on the lining of the circulatory system and then manage to get into the surrounding tissue. This process is really the same as the one they used to get away from the primary tumor and into the circulation—they're just going in the opposite direction and, as we've seen, they then use specific molecular targets as landing strips.

This metastatic cell voyage is in some ways similar to what happens after an egg is fertilized: the cell starts dividing to form a clump of cells that implants in the uterus and eventually, when it's reached a suitable size, makes its entry into the world. Having a baby is a much trickier business than one might suppose, given the world population. At least 25 percent of pregnancies end in miscarriages, and this figure rises to 75 percent with increasing age (of both the woman and the man). Despite metastasis being the most life-threatening facet of cancer, the odds against escaping tumor cells are even worse. Only about 2 percent of circulating tumor cells manage to escape so that they can form micrometastases, but only 1 percent of these manage to persist and expand into tumors. So, with a success rate of about 0.02 percent, metastasis is a mighty inefficient business. Of course, that is where the birth parallel ends: a baby is launched into a world that is waiting for it to do great things and increase the sum of human knowledge and enjoyment. The tumor cell that does manage to grow in a new home is an assassin that has taken the first steps on a journey that is probably going to kill its host.

Do Metastases Metastasize?

There's another, perhaps rather obvious, question that you might ask about metastasis. If primary tumors shed cells into the circulation and some of these eventually become secondary tumors in a new location, what's to stop cells from a metastasis doing the same thing in reverse? At least in mice, the answer is "nothing, and they do." This has given rise to the idea of what Joan Massagué of the Memorial Sloan-Kettering Cancer Center has called "tumor self-seeding" that can be visualized by marking metastatic cells (with a fluorescent label) and inoculating them into mice. The mice also have an unlabeled tumor of the same cells and it is then simply a matter of tracking the label to show that extensive seeding by circulating tumor cells is a common occurrence.

This type of metastasis may occur in at least some human cancers and, given that the cells doing the seeding have already jumped some of the major hurdles on the road to becoming a fully malignant tumor, may contribute to the aggressiveness of some cancers. There is, however, no conclusive evidence for this in humans, and it's a very difficult thing to tackle experimentally. It's an important point, though, because its resolution would have implications for treatment strategies.

HOW DOES CANCER KILL YOU?

Cancers are, of course, abnormal growths of cells and their damaging effects on the tissues they grow in can kill in a direct way. Thus, colon cancers, other tumors in the gastrointestinal tract, and also ovarian carcinomas can obstruct the bowel, which would be fatal without surgical intervention. Similarly, lung tumors can be fatal if they block lung function, and some thyroid tumors can, in effect, cause strangulation. A variant on this theme is leukemias that result in a huge excess of white cells over red cells, which so increases the viscosity of blood that circulation is drastically impaired. However, human beings are astonishingly resilient to organ damage. We can manage with half a kidney, and if we lose two-thirds of our liver, it will regenerate itself. It is therefore rare for cancers to kill via organ failure. Death usually comes from secondary effects, principally infection. Cancer patients generally become increasingly susceptible to infection owing to the decreased efficiency of their immune system. Tumors that damage the walls of tissues can also make us vulnerable to infection. Infection is commonly by bacteria (e.g., *E. coli, pseudomonas*), which can overwhelm the host even with antibiotic treatment, but also by fungi. Tumors can also cause damage to blood vessels, leading to hemorrhage, particularly in the liver.

About 40 percent of cancer deaths occur from malnutrition—a general condition of starvation and debilitation called cachexia (wasting syndrome) that develops in many other chronic diseases. Cancer cachexia is not understood, and there are no satisfactory therapeutic treatments. Both chemotherapy and radiation therapy can also induce cachexia, and weight loss in turn reduces the efficacy of chemotherapy. Metastases may also suppress the immune system, e.g., if the tumors are in the bone marrow, or, if they are in the brain, they may raise intracranial pressure.

Of the eight major changes that almost always occur as normal cells turn into tumors, the most drastic is becoming able to spread—to metastasize—throughout the body because that is usually the way in which cancers kill. Although this is a crucial development in most tumors, it is driven by a combination of changes by which cells cease to respond normally to signals regulating growth and location and avoid cell suicide, one of the main protections against cancer. The re-wiring of cell signaling systems that permits this behavior also enables cancer cells to send out signals that recruit host cells to promote their survival, including those that form new blood vessels to nourish the growing tumor. All these effects arise from mutations that affect the way signal pathways work. We now turn to DNA sequencing, the story of how the human genome was first sequenced, and how very recent developments have taken us to the point where genomes can be sequenced at very high speed, an advance with huge implications for clinical practice.

8

LET'S SEQUENCE YOUR DNA

THIS BOOK IS ABOUT CANCER: ABOUT THE GROUPS OF GENES AND proteins that cause cancers, why they make cells behave abnormally, and how we can use our knowledge of them to come up with ways of treating or even preventing cancers. It's quite possible to understand much of that without knowing anything about the intricacies of our DNA code and how it is unraveled. What you can't do without having this background, however, is begin to grasp why cancer has proved to be such a problem and why we are now in such a dramatic phase of the cancer story. Apart from that, it's a rattling good yarn—perhaps the most extraordinary in the history of science. It has already made significant impacts on medicine and it will come to have a central place in mankind's efforts to understand and control disease.

Launching one of the most astonishing periods in scientific history, in 1977 Fred Sanger and his colleagues produced the first complete sequence of a DNA genome. In 1990 the project to sequence the entire human genome was set up, and by 2003 the task was completed. In 2007 the first individual human genome sequences were published. Since then, the speed at which genomes can be sequenced has risen, and the cost has declined to the point where it is now feasible to obtain the complete sequence of someone's DNA within a day or so. Let us trace the story of the greatest scientific revolution, now in the process of changing medicine like no other.

THE MAKING OF A WORM

Like all good stories it starts with "once upon a time," though in fact the time is only about forty years ago, but the story has a very improbable hero: a worm.

The place is the Medical Research Council Laboratory of Molecular Biology in Cambridge. By 1969 Max Perutz had been the director of the LMB for seven years, and the joint heads of the Cell Biology Division were Francis Crick and Sydney Brenner. The extraordinary Brenner was a South African doctor who, with Crick, had shown in 1961 that, in the genetic code, groups of three bases of DNA determine which amino acid is added to a growing protein chain. He'd then come to the conclusion that the questions of how DNA makes protein had been solved in all but minor detail. The focus of Brenner's mind had shifted to what he saw as the greatest problem: how the DNA code makes a complete animal. That is, how to work out where every cell comes from—how different types of cell (cell lineages) emerge from a single fertilized egg to make the whole creature. He had decided that the animal for this endeavor was the worm, more specifically a nematode (roundworm) of the species called *Caenorhabditis elegans*. It had to be a multicellular organism, of course, but *C. elegans* is small (it lives in soil like any self-respecting worm, but it's only about 1 millimeter long) and, compared to the other much-used model for genetic studies, the fruit fly (*Drosophila melanogaster*), it's a pretty simple creature (worms don't have legs, wings, compound eyes, and various other complicated bits). It also has one other wonderful feature: it is transparent, so that it's possible to track every cell from the fertilized egg to the 556 cells that make a newly hatched worm and then on to the 959 cells of the adult worm. Or at least it's possible if you're John Sulston, the person Brenner recruited to help him realize his dream. Sulston had read chemistry at Cambridge and done a PhD there before going to the Salk Institute in La Jolla, California, to work with Leslie Orgel on how the very earliest forms of nucleic acids might have replicated before the evolution of complex cellular machinery. It was Francis Crick, working at the Salk as a visiting fellow, who recommended Sulston to Brenner.

In a singular saga of classical study by observation (down a microscope) Sulston, later joined by Bob Horvitz and in collaboration with Bob Waterson at Washington University in St. Louis, showed how the worm is put together. To give some idea of what a phenomenal achievement this was, the first installment dealt just with the nervous system: it defined the development of all the nerve cells (302 of them) and their 8,000 interconnections. After that they worked through each of the other cell lineages (muscle, etc.) in turn, tracing the pathways for all of the cells until they had done the entire worm.

One of the most startling findings was that 131 cells (precisely) disappear as the worm embryo develops: this is the controlled program of apoptosis that we've already seen is such an important fail-safe protection against cancer.

This showed for the first time that, counterintuitive though it may be, cell death is part of normal animal development. That was in the worm, of course, but we know now that selective killing is a general developmental feature. All this was fine in that it provided a detailed picture of how the cells of an animal developed and was of great interest to avid readers of *The Worm Breeders' Gazette*, the newsletter that circulated around the labs working on the problem, but it wasn't really what Brenner had set out to do. He wanted to understand the genetics of development: which gene controlled each step and what the encoded protein did.

By the time the worm project got underway, it was possible to identify specific genes, but it was an incredibly difficult and time-consuming business. The basic approach was to make mutants, which is fairly easy in model organisms like the fruit fly and the worm because they breed rapidly. By looking at physical patterns of inheritance, rather as Mendel had done with his peas, you can track how closely the mutations responsible are "linked"—that is, how close the genes causing the mutations are on their chromosome. This is a "genetic map," perhaps better described as a "linkage map." Note that such a "map" doesn't tell you anything about the genes themselves; they're just "black boxes" somewhere in a chromosome, and you don't have to touch DNA directly to make a genetic map.

ASSEMBLING THE BITS: CLONING DNA AND MAKING RECOMBINANT PROTEINS

The evolution from genetic maps to sequencing entire genomes was dependent on the assembly of a considerable repertoire for manipulating DNA that represented the launch of the era of genetic engineering. A critical step was the discovery of the first enzyme that could copy DNA sequences—that is, join the building blocks of DNA to produce nucleic acid polymers. The enzyme was therefore a DNA polymerase. A number of DNA polymerases have since been discovered that play slightly different roles in replication and repair, but they all make a new strand of DNA by successively joining units, the sequence being complementary to that of the template strand. The first polymerase to be isolated was obtained from the bacterium *E. coli* by Arthur Kornberg, working at Washington University in St. Louis in 1956, who shared the 1959 Nobel Prize in Physiology or Medicine with Severo Ochoa. From the 1960s on, other major advances were made, mostly in American labs but occasionally elsewhere.

Werner Arber in Switzerland began unearthing a key part of the tool kit for manipulating DNA in the form of restriction endonucleases. These are enzymes, present in bacteria and other microorganisms, that cut DNA at specific sequences (restriction sites). They've probably evolved as a defense against viruses, but that's by the by. The critical thing is that restriction enzymes can cut DNA from *any* organism; there are several thousand of them (you can buy hundreds of different ones online these days), and Daniel Nathans, Werner Arber, and Hamilton Smith won the 1978 Nobel Prize in Medicine or Physiology for finding them. Conversely, so to speak, Martin Gellert at the National Institutes of Health (NIH) in Bethesda, Maryland, discovered in 1967 an enzyme that joins DNA fragments together. So now we could cut and paste DNA.

In our first, though not last, encounter with retroviruses, we recalled the discovery of the enzyme reverse transcriptase (chapter 3). Reverse transcriptase was to become an essential part of the imminent genetic revolution because it enabled the RNA extracted from any cell to be converted into DNA. Because mRNA contains only the exons—the bits that encode the protein—the DNA sequence obtained corresponds exactly to the protein produced from the mRNA.

What was still missing was a way of making DNA fragments to order. In 1985 Kary Mullis and colleagues solved this problem by inventing the polymerase chain reaction (PCR) that enables you to make lots of copies of pieces of DNA. Now anyone could play molecular biology, and they could do it with DNA from anywhere—themselves, yeast, tomatoes, or whatever.

It's worth noting that all these tools were actually made for us by nature. In other words, all the great discoveries that set up the molecular revolution were just that: discoveries—the fruits of digging into nature, albeit with much perception allied to even greater amounts of perspiration. Indeed, the very reason that we can now manipulate DNA and genes and proteins with such exquisite selectivity is because of the specificity that comes with a biological tool kit. Key items are plasmids, small loops of DNA that occur in bacteria but are separate from the genome. They replicate independently, and you can easily persuade bacteria to take them up from the outside. So, using restriction enzymes, plasmids can be split to provide a site into which a DNA sequence of choice (e.g., for your protein) can be pasted by the joining enzyme. Persuade some bacteria to ingest your modified plasmid, give them something to eat, and let them multiply. Bacteria are very good at multiplying: Hans Kornberg, who, with Hans Krebs (he of the cycles), worked out some of their metabolic pathways, says that a bacterium has only one aim in life—to make two bacteria. This is helpful for molecular biologists because once you've taken up a plasmid, complete with your DNA insert, you can go to bed leaving them to grow for a few hours, and then extract a large number of plasmids from the bacterial culture. Now you've got a permanent stock that can be frozen and fished out whenever you want to use the DNA sequence you inserted. That process is called cloning (that's cloning a piece of DNA or an entire gene—not the same as cloning a sheep).

You can also do one other smart thing with these clones: allow the bugs to use the plasmid as a template for its transcriptional machinery. Provided you have included the right "start" signals, RNA can be made from plasmid DNA, from which the corresponding protein can be made—these are called recombinant proteins, the first being insulin.

Now you can make large amounts of either genomic DNA or of just the coding regions, and the only remaining problem is how to sequence them. We've already met Fred Sanger working out the first protein sequence, for which he won the 1958 Nobel Prize in Chemistry. The amazing Sanger won the prize again in 1980 (with Paul Berg and Walter Gilbert) for finding a way to sequence DNA. He's one of only three people to have won two science Nobel Prizes (although Linus Pauling won the Chemistry Prize in 1954 for his studies of how covalent bonds worked, particularly in proteins, and the Peace Prize in 1962 for his efforts to restrict the development of atomic weapons).

Getting at the Code

Protein sequencing remains an important technique, but it's worth noting just why being able to sequence DNA is even more crucial. First of all, if you have the complete (genomic) DNA sequence of an organism, it's possible to work out the encoded protein sequences. In addition, if you have the full sequence (as opposed to only the coding DNA) you get all the regulatory bits—promoters, etc., through which gene expression is controlled. What's more, although only about 1.5 percent of the human genome encodes proteins, we are just beginning to realize how important much of the remaining sequence is—for example, that encoding micro RNAs. So getting hold of the entire genomic sequence of an organism's DNA is really a scientific crock of gold.

Fragments of about 500 bases are ideal for sequencing using the method devised by Fred Sanger. A single strand of DNA is used as a template to make a complementary copy (i.e., the bases in the copied strand pair with those of the parent, just as happens when cells replicate their genomes). Of course, you need an enzyme to do the work (DNA polymerase) and a supply of the units (the four bases) to be incorporated in the new strand. The brilliant thing about Sanger's method, which seemed so obvious once he'd thought of it, was to add a little bit of a slightly modified form of each of the units: the modification makes it impossible for another molecule to be added to the chain. In other words, the modified unit is a "chain terminator." By juggling the relative concentrations of normal and chain terminator, it's possible to make fragments terminating at every base in the template. Now just incorporate fluorescent tags (four, giving a different light signal for each type of base) so that you can see the fragments, separate them by size, and read off the sequence.

This method of, in effect, reading the label on each base as it is tacked on to the growing chain of DNA has been used for all genomic sequencing so far. There have been some simply breathtaking advances in the speed and scale on which this can be done so that millions of DNA molecules are now read in parallel in fully automated, high throughput flow cells. These methods are called "next-generation" or "second-generation" sequencing, and their power is quite astonishing, but the underlying method remains that of measuring a light signal from a tag. Something even more awe-inspiring is on the horizon in the form of novel approaches called "third-generation" sequencing. In one version, single strands of DNA in solution are forced through small holes, the sequence coming from each of the four bases having a different effect on an electrical current flowing through the hole. Third-generation methods are still

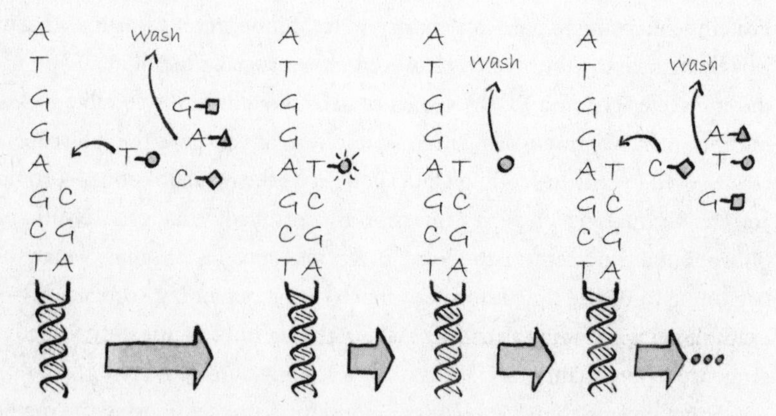

Sequencing DNA. *The sequence to be found is on the left (TCGAGGTA). This serves as a template, and the matching sequence is made by adding one base at a time. Each of the four bases is tagged with a different fluorescent label. As each is added, the light from the fluorescent label is measured, which gives the sequence.*

in their infancy, but they offer the possibility of accurate sequencing without needing to amplify or label DNA. These technical developments, among the most incredible in the history of science, have taken us to the threshold of producing complete DNA sequences within a few hours.

FROM GENETIC MAPS TO PHYSICAL MAPS

We left the worm at the stage of a genetic map while we went in search of new tools because, although genetic maps are interesting to geneticists, they're of limited use until you get your hands on the DNA itself. The first step in doing that is to a make a physical map: that is, to have clones of DNA fragments arranged in the order in which they occur in the chromosome. This became possible with the advent of cloning and restriction enzymes. We've seen that each of the enzymes in this large family recognize specific short sequences of DNA. In that sense they're a bit like the proteins that regulate transcription but, instead of controlling expression, restriction enzymes chop DNA up. It's very easy to separate the fragments: put them in an agarose gel (a polymer of sugar molecules) and apply an electric field. DNA is negatively charged so it moves through the gel, and the bigger the fragment, the slower it moves (that's called electrophoresis, and almost everyone will have seen such gels on TV programs about forensics); you can see the individual bands if you add a fluorescent dye that glows when it comes into contact with DNA. If you extract total DNA from your cells and do this experiment, the gel would just show a big smudge. That's because you have so much DNA that you get fragments of every size—a smear. But if you do it with shorter bits (e.g., plasmid clones), you will get distinct bands (and the sizes of the individual bands should add up to the total size of the plasmid). Because restriction sites are randomly distributed, different DNAs will give different band patterns. So if you have plasmids covering all the DNA in a genome, they'll give different patterns on digestion with a specific restriction enzyme (or enzymes) *unless* they share a region of DNA—that is, have overlapping segments. So, coincident patterns indicate overlapping sequence, that is, adjacent segments. Using this method, Sulston, Horvitz, Waterson, and their colleagues at LMB, MIT, and Washington University were able to build a map of the worm genome in one go, so to speak, rather than proceeding step-wise.

The worm map was complete by 1989, but you can imagine that when looking at it, you could not have avoided thinking this was just a first step: you simply had to know the full sequence of all 100 million bases. It was, as John Sulston put it, like having a school globe instead of Google Earth.

EVERY MAN IS AN ISLAND: FINDING HUMAN DISEASE GENES

Having started talking about how we use restriction enzymes, we really can't avoid a slight digression from the "worm to human to genomics" story because the way in which they were used to map the worm genome closely parallels how the first human disease genes were found. Some of those were indeed "cancer genes" but that's not the reason for spending a little longer on restriction enzymes—it's the chance to talk about what makes each human unique, part of that uniqueness being differing susceptibility to cancers.

It's widely known that DNA is the same in all human beings. Well, almost. After all, if we're about 96 percent identical to chimps and share half our genes with bananas, we sure don't have much room for maneuvering among ourselves. The "almost" amounts to 0.1 percent of our DNA (that's one base in every 1,000), which is what makes us different from one another. You may find this cheering or suicidally depressing, depending on which of your fellow men you happen to contemplate. So perhaps it's best to think of it as 3 million bases (out of our 3,000 million) that differ between us to an extent that depends on how closely related we are. With the exception of those who inherit mutations in specific genes that drastically affect their function (e.g., the one that causes cystic fibrosis), this variation evidently doesn't affect our capacity to function as humans, but it clearly does have effects—for one thing, we'd all look the same if it didn't. The biggest group of these minor DNA variants are single-nucleotide polymorphisms (SNPs, pronounced "*snips*"). SNPs are exactly what their name says: a difference of a single nucleotide (base) in two otherwise identical stretches of DNA sequence. A given SNP may be present in one allele (on one chromosome) but not in the other within one individual, or it may be in both alleles or neither. Because restriction enzymes are very particular about the DNA sequence they target, if a SNP occurs in a restriction site, it stops the enzyme from recognizing it (equally, a SNP could introduce a new restriction site). Enzyme treatment of two DNA samples (one with, one without the SNP) will thus give pieces of different size that can be separated on a gel and detected with a labeled probe (a short stretch of DNA that recognizes a specific sequence).

Because humans reproduce sexually, SNPs are subject to shuffling when we make eggs and sperm, just as happens with larger DNA segments. So, a word about sex before we get back to gene hunting and SNPs. In mammals, fertilization by a sperm gives an egg a second helping of DNA. Before fertilization, an egg nucleus contains only *one* copy of each chromosome (its haploid); when the sperm enters, it contributes its own nucleus with another set of single chromosomes. Each nucleus then replicates (doubles) its DNA, and the cell divides for the first time en route to becoming an independent organism. In humans the two embryonic cells that result from this first division contain 46 chromosomes. However, to produce those, half the DNA present in a normal cell has to be lost to make the haploid sperm and egg cells (called gametes), so that the union of sperm and egg at fertilization then restores the normal number of chromosomes. The process of halving the DNA occurs in a special sort of cell division called meiosis. A side effect of meiosis is that a bit of shuffling occurs between chromosomes (called genetic recombination), which is one of the reasons why sexual reproduction is so good (though not the only reason, obviously) because the cells that start the next generation have a unique DNA package. Clearly, when this shuffling occurs, the closer two segments of DNA are to each other in a chromosome, the more likely it is that they won't be separated and will appear "linked" in the same way in the daughter cell (the gamete). Conversely, the farther apart they are, the greater the chance that they'll be separated.

This applies equally to SNPs, but they offer one big advantage to the gene hunter: SNPs can be tracked using restriction enzymes, giving a way of finding the region of a chromosome that contains a disease-causing gene. Screening has to be carried out on a large number of individuals in the hope of finding a SNP in a known carrier of the disease that is absent in those unaffected. It is, therefore, enormously labor-intensive. It's like starting out with a phone book of genes and having no idea which one you're after. Pinning down the marker doesn't identify the gene, but it does get you on to the right page from which to refine your search. This was the method by which genes responsible for human diseases were being sought in the 1970s and 1980s, but the big snag was that it might take years to find one gene. The breast cancer–associated gene BRCA1 was located in this way, but it took two big groups over ten years.

We started talking about SNPs because they have been important in finding disease genes. However, this slight detour has two other points: (1) meeting SNPs will turn out to be useful because at least some have an impact on cancer, and (2) going, albeit superficially, through the slog of gene hunting makes you realize just how great the pressure had become for whole genome sequencing.

UPPING THE ANTE: SEQUENCING THE WORM AND LARGER ANIMALS

When we last left the worm with its complete physical map, it was indeed shouting out loud the need to sequence the entire genome. After all, making the physical map meant that in the collective freezers of LMB and St. Louis there was a clone (i.e., a stock of plasmid with the relevant bit of DNA in it) of each fragment of all the chromosomes. Just sequence them! That's what Fred Sanger, Barclay Barrell, and their colleagues at the LMB had been doing for other genomes, the first, in 1977, being a virus with a tiny amount of DNA—just over 5,000 bases. Then they did human mitochondrial DNA (about 17,000 bases). These were terrific achievements, but what we set out to do in this chapter was sequence human DNA—all 3,000 million base pairs of it. That's about 200,000 times what's in a mitochondrion and 30 times bigger even than the worm with its 100 million base pairs. So it was a problem of size and therefore of technology. It becomes a huge operation once you move from a few thousand to hundreds or thousands of millions of bases, as from the worm to human genomes. However, getting the sequence of all of your clones only gets you, to use the American vernacular, to first base. How do you work out the order of the DNA fragments you've cloned in the genome—the sequence of the sequences, if you like? Answer: make enough to get overlapping clones. Now, once you have all the clone sequences, you can use the overlaps to place them in the right order—that's called mapping.

Mapping sounds simple because in principle it is. It's a massive problem of data analysis, however: at 500 bases a go, you'd need a minimum of over 7 million clones to cover the human genome. That's ignoring the fact that no experiment is perfect so sometimes you can't interpret the sequence "reads"—sequencing is usually repeated at least ten times to obtain reliable data.

The genome sequencing story has turned out to be an extreme example of scientific achievement of any era, being heavily dependent on the methods and equipment available. As the 1980s drew to a close, two things became clear: whole genomic sequencing had to shift to an industrial scale and for that to happen, scientists would require massive computational resources to process the results. In 1985 Robert Sinsheimer of the University of California got most of the luminaries of the field together to see if there might be a kind of "team view" as to whether mapping and sequencing not just the worm but the human genome might be possible within a reasonable period. At that point, even the experts were divided between those who thought it a silly idea because the available technology was slow, and no one would put up the money for a project that might run indefinitely, and those who felt that it just might be on, given the rate of technical progress. The upshot was that the optimists won the day and even the pessimists were, mostly, persuaded. One of the decisive factors was that, even by then, people like Leroy Hood at Caltech, originally a medical doctor who'd acquired a biochemistry PhD along the way, had made big strides in developing automated DNA sequencing. This still used Fred Sanger's method but with the huge advance of a continuous readout of the fluorescent tags as each new base was added.

So this meeting really generated the momentum that was to drive whole genome sequencing. There was, it should be said, considerable debate as to whether sequencing was an academic pastime or of major importance, both within the scientific community and more generally. At an estimated cost of $3 billion, some scientists felt that the project would consume virtually all government funding and leave none for them, unless they happened to be in to sequencing. Nevertheless, in 1988 the US government set up the National Center for Human Genome Research at the NIH in Maryland with James Watson and Victor Shmerkovich as joint heads. The Human Genome Project, that is the program to obtain the complete sequence of human DNA, was launched in 1990 and the Center subsequently became the National Human Genome Research Institute (NHGRI).

From the outset, the project was to be an international effort, but it was quite clear that American scientific might would make the US the major player. The critical question from a parochial viewpoint was: What part would Britain play? At the time scientific historians would have been quietly placing their bets on "not a lot," given the well-known British track record of failing to capitalize on great ideas. Amazing to relate, just for once, they would have lost.

Because its map was complete, it seemed logical to start sequencing the worm and use that to test the water, so to speak, for the attack on the much larger human genome. As we've seen, the mapping project had been shared between the Cambridge and St. Louis groups, and it was agreed that sequencing would be similarly divided. At that point the LMB led the sequencing field, thanks to the efforts of Sanger and his colleagues. Fully aware of their capacity to make a major contribution, James Watson pulled off the considerable feat of providing NIH (i.e., American) funding for the British lab to get the ball rolling. In a unique sequence of cooperative quick-wittedness, the British government-funded Medical Research Council (MRC) provided £10 million to complete the worm sequence and followed this with money for a Human

Genome Mapping Project. It was clear that this massive project had to have its own home and the Wellcome Trust charity, recognizing the potential of obtaining the human genome sequence, invested in the project from 1992, putting up the money for a sequencing center for which they bought the Hinxton Hall estate just outside Cambridge. The new lab was named the Sanger Centre in 1993 with John Sulston as director.

The complete DNA sequence of *C. elegans* was published in 1998, all 100 million base pairs and about 20,000 genes, the first multicellular organism for which this feat was achieved. In June 2000 President Bill Clinton announced the completion of a "rough draft" of the human genome, and the sequence was largely completed by April 2003, two years earlier than planned, with approximately one third (1,000 million bases) coming from the Sanger Centre.

This prodigious achievement has required quite breathtaking developments of sequencing machines, robotics to handle clones, and computing power to process the data and make it easily usable by the scientific community. The problems that were overcome were not all scientific, however. In particular, there was a considerable battle to ensure that sequences were immediately made publicly accessible and that the data belonged to everyone—that is, genes should not be patented. That in itself was a hard-fought battle and a major triumph. Still, most stunning of all has been the rate of technical progress: the original cost estimate was $1 per base: the current (2011) figure is bases for one cent. The first sequencing machines in the early 1980s managed 10 kilo bases (10,000 bases) a day; the present rate is heading towards half a million kb per day. So the cost has gone down by a factor of fifty million and the speed has gone up by 100 million!!

This wonderful story has been most engagingly told in John Sulston and Georgina Ferry's *The Common Thread,* which explains the science with great clarity and describes many of the characters who took part. Sydney Brenner, Robert Horvitz, and John Sulston shared the 2002 Nobel Prize for their discoveries about animal development, with Brenner observing that the fourth winner was *Caenorhabditis elegans,* deserving all the honor but unable to share the monetary award.

SEQUENCING YOUR DNA

We set out to sequence your own DNA, but we've had to spend a while assembling the necessary bit and pieces. So now let's bring them together to finish the job in what turns out to be a fairly painless business. Just take a few milliliters of blood from your arm (about 1/100th of a normal blood donation); from the white cells it's about an hour's work to extract the DNA (nowadays you use a commercially available kit and just follow the instructions). Now break up your DNA into short bits (typically 500 bases—you can do that by squirting it through a hypodermic needle), stick them onto a surface, make enough copies to give a detectable signal, and then sequence. You need to do several rounds of fragmentation/sequencing to give you enough overlapping "reads" to work out the complete sequence. Now all you need is a (large) bit of computing power to piece the bits together and to work out where things are—a process called "annotation" that flags up genes, exons, introns, and all sorts of other things about the stuff that makes us what we are.

So there it is: all these wonderful discoveries and insights, made over the last forty years or so now mean that we can sequence entire genomes. We've seen something of the impact that the genomic revolution has already had on medicine, but it is not just the nature of fundamental research that is being changed forever. The experiment described above to sequence genomic DNA was slightly simplified, but nevertheless, the capacity to sequence human genomes in a few hours will soon be available at a cost that's currently dropping toward US$1,000 a go, and the implications of that technology are going to change all our lives. The major technical advance since 2003 is the replacement of Sanger's sequencing method by "next-generation" sequencing, which has resulted in the huge increase in sequencing speed.

Scientists are pretty cool dudes by and large: they don't get carried away, they are about as self-effacing a group as you could find and they don't seek the spotlight (OK, of course there are a few showmen and, whisper it, the odd charlatan, as in any walk of life, but those really are rare in science). So when the measured and conservative Nobel Prize winner David Baltimore says we are going to see miracles, and Leroy Hood talks of extending the human life span by twenty years, you can be confident you are living in remarkable scientific times.

CRACKING THE CANCER CODE

You might be wondering what all this has to do with cancer, beyond getting at tumor sequences quickly. The answer is that this extraordinary story, beginning with the study of how a small worm puts itself together, led the way to whole genome sequencing, first of the worm, then of humans, and subsequently to activity on a scale so vast that you can now look up the complete

DNA code of a huge number of animals, plants, and microbes. It founded the science of genomics, which has already had the greatest impact on medical science of any advance in our history. Indeed, perhaps only the flowering of atomic physics in the 1920s and '30s that produced the atomic bomb in 1945 could rival it as a scientific revolution.

We cannot therefore avoid returning to genomics as we look at the current state of play in cancer but, for the moment, let's take just two examples to illustrate this revolution. Following the full sequencing of the human genome, Michael Stratton, working at the Sanger Centre, completed a study that would have been inconceivable just a few years earlier—inconceivable because it required the sequence of the human genome but also because it relied on the fantastic technical developments that allowed such massive lengths of DNA to be sequenced at very high speed. Stratton selected 500 proteins (they were kinase enzymes), known or thought to be involved in cell signaling pathways that controlled growth. His group then screened by sequencing a large number of tumor samples to see if there were any mutations in the kinases. As some kinases were already known to be quite often mutated in human cancers, the fact that these appeared in some of the tumors was unsurprising. What was surprising was that one kinase gene, called BRAF, turned out to have the same single base mutation that changed just one amino acid in about two-thirds of the tumors. The tumor they had chosen was melanoma—there were over 68,000 new cases in the US in 2010 and it's the most rapidly increasing UK cancer—about which virtually nothing was known at the molecular level. Stratton's group had discovered a new "cancer gene" that played a major role in a very prevalent cancer. Within a couple of years, a drug had been produced that is very efficient at blocking the action of the mutant form of BRAF and thus offers a chemotherapeutic approach to treating melanoma. In the end, of course, BRAF's role in melanoma would have been discovered by the same sort of slow and painstaking methods that had identified the early worm genes, but that might have taken years or even decades. Whole genome sequencing and the era of genomics had made an almost immediate impact on medical science.

THE CANCER GENOMIC LANDSCAPE

The genomic revolution now means tumor DNA sequences from hundreds of patients can be compared with those of normal tissue from the same individuals. All cancers are different but nevertheless, this kind of screen has revealed characteristic mutation patterns for some types of tumor—and this is a powerful way of looking at what makes cancers tick. These patterns can be pictured as a cancer "landscape" by marking a map of each chromosome to show how often individual genes are mutated in a group of patients with the same type of tumors—a dot for rarely, a "hill" for occasionally (say in 5 percent of the tumor set), and "mountains" that flag the major cancer drivers, genes that are almost always abnormal in that form of cancer. With a little imagination you can picture such a map covering the whole genome: a landscape portrait or "signature" for those tumors. If they happened to be bowel cancers, you might guess that, when you came to the appropriate chromosome, the MYC gene would show up as a "mountain."

The capacity to carry out this kind of mutational mapping brings us for the first time face to face with the era of personalized medicine. Because cancers are such a mixed bunch genetically, it's only by having a complete mutational picture that you can begin to design therapy in a rational way. If, say, there were five "mountains" in the bowel tumors set, a cocktail of drugs that could reverse the effects of each of those mutations might be a pretty good lifesaver. At the moment, we don't have one drug that can block or reverse the effects of any of these five mutations, but at least we can now determine the targets we would like to hit.

In a little over a decade we've gone from the triumph of obtaining the complete sequence of the human genome to the era of "personalized medicine" in which it is possible to sequence the entire genome of an individual within a few days for a cost of not much more than $1,000. This has already affected the treatment and the lives of some individuals and in the years to come it will affect us all.

PART IV

DETECTION AND TREATMENT

9

WHERE ARE WE? WHERE ARE WE GOING?

WE'RE FOREVER BEING TOLD BY THE DOCTORS TO EXERCISE–WITH good reason, as we saw when we talked about how to get cancer. But what they never mention is the downside—the problem is not only convincing people to go through the slog of actually getting out of the easy chair to do the regular physical work, but the fact that folks who keep reasonably fit tend to think they're untouchable. No sporty group is more prone to this illusion than rugby players, as I can testify from frittering away much of my life on the game—when not injured, that is. Whatever level you play, it's a tough old sport, so you do get hurt, but you're never ill. There are strong parallels with American football, and hockey, come to think of it, but maybe all sports foster the notion in an athlete that he or she is immune to things ordinary folks get. In short, you're a toughie. So, when a longtime playing friend of mine was told a year or two back that he had leukemia, it came as a double shock to everyone who knew him. How could someone you have always pictured hurling themselves around the rugby field, giving all they had to their team, and being a complete pain to the opposition, be struck down by anything? But more than that, how could he get *leukemia*—a disease my generation had grown up equating with the most awful prognosis. He was always known by his nickname—Podge—which is, in the zany way of sport, as inappropriate as could be: he hadn't an ounce of spare fat and was able to play as he did because his fitness was simply alarming (to the rest of us!). Podge deteriorated rapidly and within a few weeks all of us, friends and family alike, were resigned to the worst. Of course, it seemed so unfair—he was younger than me and leukemia is not very common. How *could* it happen? All the emotions that so many experience every day as someone else discovers they have cancer tore at us as we waited for Podge to succumb.

But Podge didn't succumb. Very slowly he hauled himself back from the brink. He recovered consciousness. He started to talk again—and then to eat. And eventually he left the hospital (fantastic effort, docs!) and after a few more months, amazingly, wonderfully, he went back to work. He's gone a bit bald and he won't be throwing himself round a rugby pitch again and perhaps—probably—his leukemia will return. But that may not be for five years—or ten. For the moment, his life, and that of his family, has been returned. Twenty years ago Podge would have been dead. The fact that he's fine (well, doing OK) is a result of the tremendous advances made since then that mean some cancers, once killers, can now be stopped in their tracks.

It's now sixty years since the first drugs began to be used as a complement to surgery and radiotherapy for the treatment of cancers. The intervening period has seen stunning triumphs with remission rates for some cancers that were formerly untreatable now at nearly 100 percent. Moreover, we now have highly effective vaccines against some DNA tumor viruses. Innovative techniques are making tumors easier to detect, and markers that can be isolated from blood are increasingly promising for revealing the earliest stages of cancer and for monitoring response to therapy. Most dramatically of all, whole genome sequencing has already begun to change how we diagnose, classify, and treat cancers and, through the detection of yet more mutations, is increasing the range of potential drug targets.

In this cancer journey we've seen how the contortions of mutated molecules change the behavior of cells so that, rarely and almost incredibly, tumor cells emerge with their extraordinary capacity to enlist help from their surroundings with the result that they can survive and spread. In this final chapter of our story, we come to what is both the most important and the most exciting part. Important because it's about what we now have in our armory to deal with cancers that can't be cured by surgery. Exciting because the speed of developments is quite breathtaking. The history trip (chapter 1) charted the major advances in surgery that, by the beginning of the twentieth century, had really set up the surgical basis for cancer treatment. At that time, for a cancer to be curable it had to be sufficiently small and localized that it could be cut out, and this remains the first choice treatment for most cancers—so long as they haven't spread. Refinements have come with time so that these days only very small amounts of tissue are generally removed and the success rate is high; that is, tumor regrowth is rare.

In addition to surgery, radiation treatment has been progressively introduced as a way of destroying small tumor growths missed during surgery. Radiotherapy uses highly focused beams of X-rays or neutrons although it is, of course, not specific in that it will kill normal cells just as readily as tumor cells. Here too improvements over the years now give much better control, so that side effects are limited and radiation is relatively well tolerated.

These advances have been important, but the great cancer drama of the last fifty years or so has been the gradual addition of drugs to these mainstays of cancer treatment. This is what's called chemotherapy, the idea being to destroy tumor cells that have spread beyond the reach of the surgeon and the radiotherapist. Treatment carried out *after* an operation to reduce the risk of tumor recurrence is called adjuvant therapy, which includes radiation therapy and drugs. Adjuvant therapy was first shown to be effective in breast cancer, but in various forms it's now used for most cancers. Sometimes drugs are used to reduce the size of primary tumors *before* surgery or radiation treatment, and that's called neo-adjuvant chemotherapy. For most cancers, drugs are used together with radiation therapy (chemoradiotherapy or chemoradiation) or surgery (combined modality chemotherapy).

The idea behind all these treatments is, of course, to cure cancer—to prevent it from reappearing at the original primary site and to stop secondary tumors from forming. When that seems impossible, drugs are sometimes used to alleviate symptoms and perhaps give some extension of life; this is called palliative chemotherapy.

We'll look now at some of the great strides that have been made in chemotherapy, and then we'll check out some of the ingenious ways in which tumors can be detected and their response monitored once treatment begins. Focusing on drugs means that we can't avoid some fairly clunky names. However, such is the profile of cancer nowadays that a lot of these will be familiar, though there is an extra complication in that many have both a brand name and a different chemical name. For the tongue twisters, I tend to go for the first syllable, much as one does with friends—especially if they're Russian.

MAKING DRUGS FOR CANCER

In the most general sense, chemotherapy means the chemical treatment of disease—the same sort of treatment that the Egyptians were attempting with arsenic. It also covers the more modern example of taking penicillin for a bacterial infection, when you would indeed be undergoing a course of chemotherapy. However, the term generally means using chemicals to treat cancer, the agents being taken either singly or in combination, with the aim of killing tumor cells. The drugs are normally administered systemically, meaning that they circulate in the bloodstream and can, in principle, affect every cell in the body. With very few exceptions, chemotherapy drugs are not specific though there may be some targeting if, say, you hit the cell cycle of rapidly dividing tumor cells. However, and perhaps slightly counterintuitively, most cancer cells proliferate slowly relative to many normal cells and, moreover, their division rates vary widely between different types of tumor. This problem has focused much effort on producing toxic agents directed more specifically to tumor cells, either through binding to proteins that are exclusively, or at least mainly, present on tumor cells or by targeting a signaling pathway that's a strong driver. Hormone therapy, targeted therapy, and gene therapy are essentially variants of chemotherapy that attempt to enhance targeting to tumor cells.

We've already mentioned one of the earliest demonstrations that chemicals could affect cancers in the work in the early 1940s of Charles Huggins, who showed that the growth of prostate tumors could be controlled by hormonal treatment. The ensuing twenty years saw the emergence of the first drugs to be effective against various types of leukemia. A major contribution was made by the Buffalo-born pathologist Sidney Farber, working at the Harvard Medical School in 1948, who showed that a drug called aminopterin could give temporary remission of acute lymphoblastic leukemia (ALL) in children. In the 1950s aminopterin was supplanted by more effective agents—methotrexate, 6-thioguanine, and 6-mercaptopurine, shortly followed by 5-fluorouracil for solid tumors. Each of these drugs stop DNA from being made in any cell that's trying to do it, so they're not specific, even though they have been useful.

It was not until the 1960s, however, that the now common practice of using combinations (cocktails) of drugs was first tested in humans. The results were spectacular in showing that several agents acting simultaneously could be much more effective than any of them alone. The use of vincristine, methotrexate, 6-mercaptopurine, and prednisone lifted the remission rate for childhood ALL from essentially zero to 60 percent. By 1996 further advances in combination therapy had raised the five-year survival rate to 81 percent. Children with acute myeloid leukemia (AML) now have survival rates between 50 percent and 70 percent, and over 80 percent of children with acute promyelocytic leukemia (APL), a subtype of AML, are cured. By 1980, combination strategies had made similar impacts on Hodgkin's disease and testicular cancer. These days, provided they are detected early enough, most survive these cancers. The idea is that you combine inhibition of cell division and the immune response with a blockade of DNA synthesis—a triple hit that gives tumor cells little option but to throw in the towel.

Combinations of drugs are often used in a "chemotherapy regimen," the idea being to hit different targets in the cancer cell, thereby increasing the "fractional kill"—the proportion of tumor cells eliminated. Chemotherapy regimens are often denoted by acronyms, some of which are fairly logical: e.g., FOLFOX: folinic acid (FOL), fluorouracil (F), and oxaliplatin (OX), used to treat bowel cancer, and FOLFIRI: folinic acid, fluorouracil and irinotecan (sometimes also with cetuximab), also used for bowel cancer. Others, however, are pretty gnomic (e.g., .Stanford V), and it is fortunate that these days one can refer to http://www.chemocare.com/bio/list_by_acronym.asp and to the CRUK site (http://www.cancerhelp.org.uk/default.asp) for enlightenment about any drug or combination of drugs that have been prescribed for oneself or a loved one. (While we're talking about advances in health care and disease treatment, let's not neglect to at least note the massive advances in information availability—and therefore a patient's power to ask questions about and advocate for his or her own therapy options—that is, the Internet. See Sources and Resources.)

What we'll do next is review the range of drugs that have emerged since the 1980s before looking at how good they've been. Then we'll turn to the vital matter of cancer detection and how therapy may evolve as we move through the twenty-first century.

A Big Turn-off: Kinase Inhibitors

We've seen that one of the most powerful cancer drivers is a mutation that hyperactivates a signal pathway and one of the best ways of doing that is to have a mutant receptor. A lot of effort has therefore gone into drugs that block receptors; if they hit a mutant form (i.e., on tumor cells) rather than the normal, so much the better. Because the active receptors are enzymes that transfer phosphates, that step is the obvious one to target, and indeed most drugs of this sort do just that (they block ATP reacting to give up a phosphate). Given that there's a big family of receptors and they have very similar 3D shapes, it's pretty amazing that there are inhibitors that can stop or shrink tumors without killing the patient through their side effects. However, proteins are a bit like the fair sex in that even close relatives differ in the exquisite detail of their shapes, and for this reason some small molecule kinase inhibitors are clinically useful. They work essentially by interposing themselves between ATP and its binding site on the receptor.

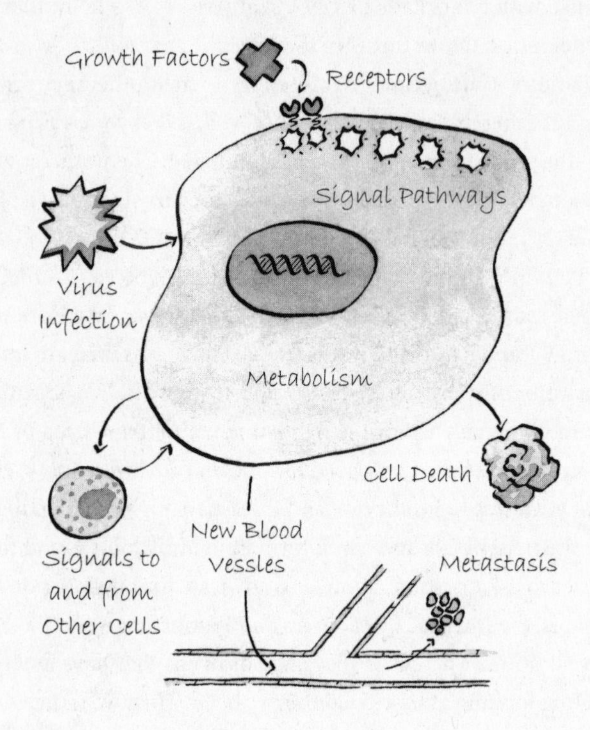

Potential targets for anti-cancer drugs. *Growth factors. Receptors. Signal pathways. Cell death. Metabolism. New blood vessel formation. Metastasis. Signals to and from other cells. Infection by oncogenic viruses.*

Eleven kinase inhibitors currently have FDA approval with a further eighty in clinical trials. Most directly block ATP binding, thereby cutting off the source of phosphate, and it's perhaps slightly surprising that, despite targeting a common feature, these agents have some specificity for individual kinases. It's also a bit surprising that these drugs are sufficiently well tolerated to make them usable, bearing in mind that most of them can't distinguish between tumor cells and normal cells. One exception is imatinib (Gleevec), used for the treatment of chronic myelogenous leukemia (CML), because it targets the mutant BCR-ABL1 protein that is expressed only in the tumor cells (chapter 5). This drug has had the wonderful effect of giving an eight-year survival for a condition that hitherto had been uniformly fatal. Nevertheless, the story of Gleevec is a cautionary tale in that, although most CMLs initially respond to the drug, small clones of cells with a mutated version of BCR-ABL1 emerge so that they eventually become dominant and confer resistance. To counter this, a second-generation inhibitor, dasatinib, can be brought to bear that actually works quite well, even though it's less specific than Gleevec—that is, it blocks a number of other kinases as well as BCR-ABL.

Selective Estrogen Receptor Modulators (SERMs)

We all know about steroid hormones, if only from the sports news, but they're also much to the fore in the context of breast cancer where the key hormone in question is estrogen. Much effort has been directed to controlling its activity, particularly in the form of drugs called selective estrogen receptor modulators (SERMs). These stick to the hormone receptors and either mimic the action of the natural hormone or prevent it from acting. SERMs are "selective" because some activate the estrogen receptor in all cell types (estrogen), some block it in all tissues (fulvestrant), and some can have either effect depending on where they act (tamoxifen). If you're asking "Why the different effects?," well done! It's a very good question and, as we have only a hand-wavy answer, we'll ignore it, merely pointing out that this sort of behavior shows just how tricky the drug business is.

Since 1980, tamoxifen—maybe the best known of all anti-cancer drugs—has been the standard anti-estrogen therapy for breast cancer. It's what's called a pro-drug (meaning that it's converted to its active form by the body after administration). The problem of different effects depending on where you look is illustrated by the fact that in the uterus it activates the estrogen receptor to *switch on* its target genes. In breast cells, however, estrogen receptor/tamoxifen complexes bind to DNA perfectly well, but now they *block* activation of the genes that estrogen would normally turn on, so the effect is to stop cell growth. Other names that are becoming as familiar as tamoxifen are aromatase inhibitors (anastrozole, letrozole). Instead of blocking estrogen, they block aromatase, which carries out one of the steps in making estrogen from cholesterol so that the result—stopping the flow of estrogen that can exacerbate breast cancer—is the same.

"Magic Bullets": Monoclonal Antibodies

The idea that you might be able to make something with selective killing power originated with Paul Ehrlich and led in 1910 to Salvarsan, a drug that targets the bacterium that causes syphilis. He also came up with the word "chemotherapy," but full realization of his concept of designing molecules that were super-specific for a target had to wait until the 1970s when technological reality began to catch up to Ehrlich's imagination, and we were able to create a method for making monoclonal antibodies. Antibodies are large proteins made by the immune system, but their critical bit is a small region that locks tightly onto the foreign target. In essence, making a "monoclonal" is straightforward: inject the protein you want it to recognize into a mouse, wait a while, and then collect lymphocytes from the mouse that by then are making antibodies that stick to your protein. Fuse the lymphocytes with an immortal cell line that doesn't make antibodies (another use for anti-freeze) and you should be able to isolate a clone of cells you can keep on growing that makes only one type of antibody (i.e., a monoclonal). The antibodies are mouse proteins so they can provoke an immune response in humans. To get around this problem, you can use DNA engineering to make "humanized antibodies" that, in effect, have the critical binding bit (mouse) in an otherwise human antibody—so these should give exquisitely selective binding (because they actually use what the mouse has come up with to defend itself) with no immune reaction (because they're almost entirely human).

The first cancer "monoclonals" were rituximab (Rituxan) and trastuzumab (Herceptin). These drugs bind to two receptors (CD20 and the ERBB2, respectively), and received FDA approval in the late 1990s. CD20 is a protein on the surface of some white blood cells, and using rituximab as an intervention had an immediate impact on the treatment of non-Hodgkin lymphomas that were resistant to other drugs. Herceptin is used, in combination with conventional chemotherapy, to treat metastatic breast cancers that express ERBB2. Although it's been effective, a high proportion of tumors are unresponsive, even those that make ERBB2, and those that do respond generally develop resistance.

Several other monoclonal antibodies now have FDA approval for cancer use, notably cetuximab (anti-EGFR) for bowel and head and neck cancers. Monoclonals have also been used in three different ways to deliver toxic agents to tumor cells: (1) by linking antibodies against proteins on tumor cells to radioactive isotopes, so that the target cell is killed by radiation (radioimmunotherapy); (2) by linking antibodies to an enzyme that activates a pro-drug—when a pro-drug is administered, it's activated as a killer only near the cells that have been targeted (called antibody-directed enzyme pro-drug therapy, or ADEPT); and (3) by making artificial lipid sacs containing drugs or radioisotopes so that by incorporating antibodies in the membrane these can be directed to tumor cells (immunoliposomes). Despite the ingenuity of these methods, in general, attempts to target toxic agents to tumor cells have been unsuccessful.

Cutting Off Supplies: Targeting Tumor Blood Vessels

An alternative approach to trying to knock out tumor cells is to turn off their blood supply. In 2004 the humanized version of a monoclonal antibody, bevacizumab, better known as Avastin, became the first FDA-approved anti-angiogenic drug. Avastin can be thought of as a drug that shoots the messenger because it sticks to VEGF (vascular endothelial growth factor), the most potent inducer of new blood vessels, and blocks its receptor interaction. Initial trials showed that Avastin alone didn't do very much, but it had significant effects when combined with other drugs. It was therefore initially approved for metastatic bowel cancer in combination with 5-fluorouracil. Subsequently its range has been enlarged to include some types of lung, breast, and brain tumors, and it is currently undergoing trials for use against melanoma as well as cancers of the ovary, kidney, stomach, and prostate. Despite its promise, Avastin received a setback in 2010 when the FDA rescinded its approval for its use against breast cancer because cumulative evidence has shown that it does not prolong life but does cause serious side effects. Avastin retains approval for the treatment of other cancers.

Another category of drugs targets the vessels of the circulatory system directly by hitting the endothelial cells that line the vessels, which can lead to tumor nutrient starvation. These are not at all specific because they disrupt proteins involved in pulling cells apart when they divide, so that any cell attempting to replicate itself is affected. However, they have been widely used in the treatment of cancer, particularly the category known as taxanes, and may be successful because in the blood they "see" only the lining cells, most of which are not dividing unless they are part of a tumor.

Gene Therapy

If vital genes are damaged or lost from cells, the most obvious thing to do is to put back a normal version. An alternative for cancers is to put into tumor cells a gene that will kill them. All such gene insertion methods come under the heading of "gene therapy," but so far they have been almost impossible to realize for the treatment of cancer. The most common approach uses viruses with the therapeutic gene put into a viral genome that has been modified so the virus itself is not harmful. This method then relies on the infectious capacity of viruses to carry the gene into cells, following which the encoded protein is made. Viruses, of course, do not infect only tumor cells, and targeting remains a problem, although it is possible to engineer viruses to express proteins that bind to specific cell surface receptors. The first gene-replacement therapy trial took place in 1990 when lymphocytes from a four-year-old girl with severe combined immunodeficiency disease were exposed outside her body to viruses containing a normal version of the gene that's defective in this condition. Injection of the engineered cells allowed her immune system to begin functioning. The problem with cancers, of course, is that you can't remove the cancer cells from the body to treat them; if you could, you wouldn't need to worry about therapy of any sort.

There is one elegant example of the viral strategy in a modified virus that can destroy tumor cells that have lost the P53 tumor suppressor gene while having no significant effect on normal cells. This illustrates the enormous promise of gene therapy, but technical difficulties have meant that progress thus far has been slow.

The Holy Grail: Vaccines

The dream of being able to take something that would keep us from ever getting cancer has been around for a long time so we should perhaps dispose of the subject. We've seen that cancers are a multicomponent consequence of being alive and, because of the limitless ways in which they can arise, there is never going to be a general cancer "vaccine" of the sort that works against diseases caused by a single event—say, an infection of some sort. Nevertheless, the term "vaccine" is used in two cancer contexts, the first being prophylactic vaccines. These are essentially traditional vaccines because they *do* protect against infection by a specific agent by activating an immune response, the antibodies thus produced blocking infection. About 5 percent of *all* cancers worldwide are caused by human papillomaviruses (HPVs), and one of the great triumphs of science in the battle against cancer has been the development of vaccines (Cervarix and Gardasil) that appear to give virtually complete protection against infection by the tumor-promoting HPVs. The vaccines are artificial virus-like particles that are non-infectious but induce a strong immune response. The antibodies thus produced block binding of HPV to their normal target cells and hence prevent infection. A variety of hepatitis B vaccines are also now available that work in a similar way by blocking initial viral infection; these too appear to confer sustained and complete protection. These vaccines, however, work only to *prevent* infection: they're of no use once infection has occurred.

The second type are therapeutic vaccines where the idea is to boost the immune response of the patient; they can therefore be used not only to inhibit the growth of tumors but to complement conventional therapies. In principle, vaccination with protein fragments that switch on the immune system is the simplest method for enhancing natural defenses against tumor development. The fragments used are from proteins expressed specifically on the surface of tumor cells, a considerable number of which have now been identified. Several such vaccines against various leukemias, breast cancer, and bowel cancer have been tested on small numbers of patients. Although progress in this field has been slow, Sipuleucel-T (Provenge) was approved by the FDA in 2010 for prostate cancer. The immune response it provokes has prolonged the life of patients with the advanced stage of hormone-resistant prostate cancer.

Despite some positive results, in general, therapeutic vaccines have been disappointing and the focus of research has shifted to what's called adoptive cell transfer (ACT) therapy. In ACT, lymphocytes that have invaded a tumor and can attach to proteins on the cancer cells are removed and grown in the lab to make large numbers of cells that react against the tumor. The patient then has his lymphocytes knocked out by chemotherapy before the "activated" cells are infused as a replacement. ACT has been particularly effective in patients with metastatic melanoma. However, overall progress in this field of therapeutic vaccines has been slow, and no cancer treatment vaccine has yet been approved by the FDA.

THE PERVERSITY OF LIFE: DRUG RESISTANCE

The main reason that drug treatments for cancer either don't work or gradually lose their effectiveness is that tumor cells become resistant to their effects. This is a very variable effect. Some tumors have generally low levels of resistance (e.g., Hodgkin lymphoma and childhood acute leukemia) whereas others usually respond to initial treatment but eventually acquire resistance (e.g., lung cancers), and some types are inherently resistant (e.g., melanoma). Resistance mainly comes about because the target acquires new mutation(s) so that the drug no longer works, or more simply because cells make protein pumps that shunt the drug back out of the cell. These pumps appear to have evolved as rather general cell rubbish shifters and just turn out to be quite good at carrying anti-cancer agents.

A further problem is that among the nonspecific targets of drugs will be cells of the immune system, so that one of the innate defense systems may be affected by agents designed to kill tumor cells. One counter to this is to combine immunotherapy—the use of a vaccine or antibodies to boost the immune system against tumors—with conventional chemotherapy drugs to give a more effective anti-tumor response than either method on its own can manage.

HOW WELL DO DRUG TREATMENTS WORK?

Despite these problems, the five-year survival rates for a number of cancers have increased significantly over the latter part of the twentieth century, as we noted in chapters 2 and 3. The overall five-year rate for white Americans diagnosed between 1996 and 2004 with breast cancer was 91 percent. For prostate cancer, non-Hodgkin lymphoma, and leukemia the figures were 99 percent, 66 percent, and 52 percent, respectively. This reflects a long-term trend of increasingly effective cancer treatment, and there's no doubt that the advances in chemotherapy summarized above have contributed. However, just how useful drug treatments have been remains controversial and impossible to disentangle from other significant factors, notably earlier detection and improved surgical and radiological methods. On the cautionary side, we should also note that for metastatic cancers there has been little change in the survival rates and that for breast cancer, for example, although chemotherapeutic advances have reduced the rate of recurrence, it still does recur in a substantial proportion of cases.

SEEING TUMORS

The diagnosis of cancer has historically relied on clinical examination together with imaging methods and looking directly at tumor samples by microscopy. Imaging mainly means using X-rays to form 2D images. It also includes scintigraphy, which uses radioactively labeled agents to give 2D maps, and thermography, used particularly for mapping surface temperature increases associated with breast tumors. In addition, ultrasound imaging has been used, particularly for the detection of breast and prostate cancers. Medical ultrasonography uses acoustic energy in the form of sound waves at a frequency above the normal audible range (this is the ultrasound detection method that enables mums and dads to watch junior in action before he enters the world)—as a side note, one of its inventors was Carl Hellmuth Hertz, the son of the nephew of the aforementioned Heinrich Hertz.

Positron Emission Tomography (PET)

Aside from conventional X-rays, the cuddly sounding method of PET (positron emission tomography) is widely known as a way of finding tumors and following their response to treatment, and it's also used to investigate a range of neurological disorders—things like Alzheimer's disease. It's usually done by injecting a form of glucose (deoxyglucose) with an attached radioisotope (fluorine–18) into the bloodstream. Because of their weird metabolism, noted by Warburg all those years ago, as fluorine–18-deoxyglucose (FDG) circulates around the body, most tumor cells take it up more avidly than do normal cells, and that shows up in a computer-generated image or what is more commonly known as a "PET scan." The method was first devised in the 1950s by Gordon Brownell, Charles Burnham, and colleagues at the University of Pennsylvania and Massachusetts General Hospital (Plate 10).

Because, after injection, FDG becomes trapped in any cell that sequesters it, tissues with high glucose uptake, such as the brain, the liver, and many cancers, give a strong signal. Slightly unhelpfully, not all cancer cells take up significant amounts of FDG, but PET is commonly used to detect Hodgkin's disease, non-Hodgkin lymphoma, lung, pancreatic, and esophageal cancers.

Single photon-emission computed tomography (SPECT) is similar to PET although the radioisotope is usually attached to a molecule that acts like a messenger and targets a cell surface receptor. PET or SPECT can be used together with computed tomography (CT) to provide one of the most powerful methods for precisely locating tumors. CT uses X-rays to acquire two-dimensional images; from a large number of such images, taken as the radiation beam moves through the body, a three-dimensional picture can be pieced together. Because this permits whole organs to be visualized, it has become an immensely powerful diagnostic tool since its introduction in the early 1970s. PET/SPECT-CT therefore combines the metabolic activity signal from PET with the anatomical image generated by CT.

Although PET and CT scanning are considered to be noninvasive, they still involve ionizing radiation. We observed in chapter 2 that a typical radiation dose from a CT scan is several hundred times that from a chest X-ray and, while the risk of such an exposure leading to cancer is small, it is not zero. In the US, where one million CT scans are carried out each year, it is estimated that about 1.5 percent of all cancers might be attributable to an individual's having had one single CT scan. As there are 2 billion scans worldwide, the annual contribution to the cancer burden is not negligible although the risk has, of course, to be weighed against the benefit. The situation thus contrasts with, for example, smoking one cigarette, estimated by life insurance companies to shorten life expectancy by ten minutes.

Magnetic Resonance Imaging (MRI)

The earliest report that human tumors could be distinguished from normal tissue by nuclear magnetic resonance (NMR) emerged in the 1970s, and by the end of that decade the first whole-body scans had been carried out. From these beginnings, magnetic resonance imaging (MRI) has evolved to become a widely used radiological method to construct two- or three-dimensional images of the body and hence identify structural abnormalities. MRI usually images the hydrogen atoms (protons) in the water content of the body using a strong magnetic field that aligns atomic nuclear spins, followed by a short radio wave pulse that "flips" the spin of the aligned nuclei. After "flipping," protons return to their equilibrium state at different rates depending on their surroundings, so that different tissues can be identified. A typical MRI scan consists of about twenty sequences and, depending on how the data are analyzed, the images show water and fluid within tissues as dark regions with fatty tissues being light or, alternatively, highlight water. Compared to PET, MRI gives greatly enhanced soft-tissue contrast and resolution—particularly valuable in delineating tumors. In contrast to PET and CT, NMR methods do not use ionizing radiation, and they are considered to be noninvasive and very safe.

What is more, images can be obtained of slices of the body in essentially any orientation through a tissue. This has led to the development of 3D scanning methods, giving MRI a valuable role in delineating solid tumors prior to surgery (Plate 10).

In addition to anatomical imaging, NMR signals can reveal the presence of specific chemicals in tissues. Broadly speaking, NMR has shown that the levels of lipids may reflect the rate of cell proliferation, metastatic capacity, and the acquisition of drug resistance. Proton NMR has so far, however, been relatively limited in both its diagnostic capacity and in monitoring early tumor responses to drugs.

Despite these limitations, the rapidly developing field of tumor imaging is of potentially great importance in both detection and monitoring response to treatment. The combination of MRI (or CT) scans with PET images (called image fusion or co-registration) is used to produce the most refined tumor maps. This has led to the design of programs that precisely match the beams of therapeutic radiation to the contours of the tumor so as to minimize exposure of healthy tissue. Together with developments in radiation methods, these advances mean that screening and diagnostic imagining technology as well as radiotherapy is steadily improving in terms of dose refinement and radiation targeting.

Tumor Flags

Usually called biomarkers, tumor flags are proteins or small molecules in the blood that come from tumors. If you can detect them in the blood, you know there's a tumor somewhere in the body, and if they change in concentration in response to a treatment, you have a monitor for how well that treatment is doing. The leading candidates for biomarkers are proteins released into the circulation from tumors, one much in the news being prostate specific antigen (PSA). The detection of elevated blood levels of PSA would prompt the recommendation of a prostate biopsy. However, 7 in 10 men with raised PSA levels will not have cancer, and more than 1 in 5 of those with prostate cancer will have normal circulating levels of PSA.

Ovarian cancer cells quite often release something called cancer antigen 125 (CA125). Blood levels of CA125 are raised in about 50 percent of these cancers at an early stage and in about 90 percent of advanced cases. Decrease in the circulating level of CA125 has been associated with a positive response of ovarian cancers to treatment.

These examples illustrate the current predicament over cancer biomarkers. The PSA test has FDA approval, but the high false positive and false negative rates indicate its limitations. Similarly, although CA125 is used to monitor the response of ovarian cancers to treatment, it is severely limited as a diagnostic assay. Much effort is going into this problem, but at the moment it has to be admitted that there are no biomarkers for cancer that approach an ideal specification.

It hardly requires saying that the entire field of cancer therapy is one of the most rapidly moving in science. New drugs, novel methods of drug delivery, and more sophisticated methods for tumor detection and monitoring are continually under development. A range of radiolabels is being investigated to improve the sensitivity of PET scans, and so-called smart contrast agents are coming along that will improve the resolution of MRI images. Inert capsules, nanotubes, graphene, nanocells, and nanoparticles are some of the innovative methods being developed to improve drug delivery, although it has to be said that, for tumors in general, targeting remains an unsolved problem. When we sequenced your DNA, we noted how the term "genomics" emerged from sequencing the worm and human genomes. From this has developed a range of "omics," in particular, proteomics (meaning, by analogy with genomics, the study of proteins), transcriptomics (RNA transcription from DNA), and metabolomics (the analysis of the metabolites present in a biological sample, cell, tissue, or fluid at a given time). The latter can give a "metabolic profile"—a "chemical snapshot" of a sample (urine or blood), revealing what your body was making at the time. In terms of tumor biomarkers. optimism that things will improve largely rests with the emerging field of metabolomics.

WHICH GENES ARE "ON": GENE EXPRESSION PROFILING

The rate of advance of chemotherapeutics in the twentieth century was in part limited by having as targets only the genes and their proteins that had been found by the slow and laborious methods then available. It's been possible since the 1970s to see whether a particular gene is "on" (i.e., making RNA from a sample) but you could find this out only one gene at a time. The field was revolutionized by the arrival of DNA microarrays in which 10,000 short DNA "probes," each recognizing a different gene, were stuck on a small glass slide. Now it was possible to obtain a genome-wide picture of which genes were turned on, and "expression profiling" began to be applied to tumor samples at the end of the 1990s.

An early striking result showed that two types of acute leukemia (ALL and AML) could be distinguished from the gene expression patterns, quite independently of the usual method of diagnosis, which was looking at blood samples. Other influential studies followed, notably the identification of a seventy-gene "expression signature" strongly associated with the formation of metastases from primary breast tumors. This provided a much more robust indicator than had previously been available from studies based on one or a few genes. The results had an immediate impact on treatment by showing that about three-quarters of patients with no sign of disease in their lymph nodes and who would, therefore, have been candidates for adjuvant therapy, would *not* have gone on to develop distant metastases. That is, they would have been given a treatment from which they could not have benefited and from which they might well have suffered side effects. The seventy gene "poor prognosis" signature has subsequently been developed as a molecular diagnostic test (MammaPrint) for breast cancer.

WHOLE GENOME SEQUENCING IS GOOD FOR YOU

The word "genome" was first coined in 1920 as a blend of "gene" and "chromosome." The study of genomes, that is, of the DNA sequences of organisms, emerged as genomics in 1980 with the achievement by Fred Sanger and his colleagues in obtaining the first whole genome sequences. In the last chapter we saw that since 2003 dramatic technical advances in the shape of "next-generation" sequencing have produced a phenomenal increase in the rate at which sequences can be obtained. This has ushered in the era of "personalized medicine," meaning that individual genomes can be sequenced in a day for a cost approaching US$1,000. This has the revolutionary implication that the mutation pattern of an individual tumor can be used to design a therapeutic strategy *before* treatment is started. It will also be evident that "genetic profiling" offers the best screening method for early detection—that is, a mutation pattern characteristic of a cancer may appear long before there are any clinical signs (palpable lumps, bleeding, etc.). From this it is evident that the science of genomics is poised to make the greatest impact on medical science in human history. The following examples illustrate some of the major advances that have already occurred in the first decade of the twenty-first century.

It should be said that the huge effort that has gone into sequencing has not been without its critics and, to make clear which side of the fence we are on, we will stress the huge advances in cancer that it has already brought. Before we get to that, however, we might recount the story of a young Wisconsin boy admitted to the hospital with an inflamed bowel that, as the cause continued to elude the best efforts of his physicians, deteriorated until it seemed that he would die. As a last resort, someone suggested sequencing his DNA. This revealed that he had inherited a novel mutation that caused a defect in his white blood cells. This was treatable, and the boy recovered fully.

WHAT WHOLE GENOME SEQUENCING (WGS) TELLS US ABOUT CANCER

A large number of blood samples were collected for the Human Genome Project. Just a few of these were pooled, and DNA from the white cells provided an anonymous source for sequencing. In 2007 the J. Craig Venter Institute released the first individual genome sequence, that of the institute's founder. This was followed almost immediately by that of James Watson (he of DNA), the first genome to be sequenced using "next-generation" rapid-sequencing technology. These were "proof-of-principle" efforts, so to speak, and they led to the simultaneous publication in 2008 of the first three complete sequences of individual human genomes. Two of these were from a male Nigerian and a male Chinese. From a cancer point of view, however, the really interesting one was from a female who had died from acute myeloid leukemia (AML) and, because Rick Wilson's group at the Genome Center at Washington University obtained the complete sequence from both normal and tumor tissue, it was possible to look for differences. Astonishingly, it emerged that the tumor cells had acquired over 30,000 single base changes. The amino acid sequence of ten proteins had been changed by these mutations, and if that seems rather a small number, recall that most of our genome is non-coding and that non-coding SNPs (single-nucleotide polymorphisms) may also affect gene function.

Since then, genomic sequencing of substantial numbers of patients with a range of cancers has gathered pace. At one level, from what we already know about cancers, the results have been predictable: a huge number of mutations accumulate in the genomes of tumor cells. Even so, the variety of mutations has surprised seasoned molecular biologists and highlighted the flexibility of the human genome. As in the AML example, tumors accumulate a large number of SNPs, some of which affect protein sequences. In addition, some stretches of the genome are lost, and any genes therein have, in effect, ceased to exist as far as the tumor is concerned. Other regions are amplified, giving multiple copies of genes, with potentially oncogenic effects (recall that an oncogene is simply a gene that has the potential to cause cancer).

A major surprise, however, has been the discovery that gene rearrangements, known to be common in leukemias, also play an important role in other cancers. One set of primary breast tumors had an average of thirty-eight such rearrangements per tumor with over 200 present in some tumors. Most frequently, these mutations produced repeated, adjacent, identical sequence blocks of DNA (called tandem duplications) and, excitingly from a diagnostic viewpoint, different subtypes of breast cancer have different numbers of duplications.

Up to the point where it became possible to sequence the entire genome, it had been assumed that genes came in pairs. That is, everyone had two copies of most genes, one on each of the chromosome pairs (apart from the X and Y chromosomes). Quite unexpectedly, once we could look at the complete sequence, it emerged that, although we do indeed have two copies of most genes, for about eighty of them the number varies among us. Mutations in some tumors amplify this effect, giving large-scale copy number gain. These are massive genomic perturbations: the regions involved can be hundreds of mega-bases long and they occur across more than half of the entire genome with the copy number of some genes being increased by more than ten times.

Taken together, these discoveries from WGS of individual tumors illustrate the bewildering complexity of instability in cancer genomes, only now being fully unveiled through the power of next-generation sequencing. From the application of WGS to a variety of cancers, it is already clear that the method can give three key pieces of information: (1) reveal new mutations; (2) identify core signaling pathways that are almost always abnormal in a given type of tumor, even though the mutated gene(s) differs between tumors; and (3) reveal mutational signatures permitting the subclassification of tumors, which can help in prognosis and in the design of more specific treatments.

As this approach is extended to all the major cancers, it should greatly increase the range of diagnostic and predictive biomarkers. We have already mentioned how genomics can provide biomarkers through the application of gene expression profiling to breast cancer. However, WGS has advantages over expression profiling in that it is easier to obtain DNA than RNA, and the complete sequence reveals all mutations. This includes amplifications, translocations, tandem duplications, and copy number variations, the extent of which is only just beginning to be recognized.

SCREENING WHOLE GENOMES TO PREDICT CANCER RISK

About 10 to 15 percent of cancers are hereditary, depending on the type of cancer. For example, approximately 10 percent of breast cancers and 35 percent of bowel cancers arise from genetic abnormalities present at birth. BRCA1 and BRCA2 are two well-known major susceptibility genes for breast cancer, inherited mutations in either conferring a high risk of developing the disease.

Variants in some other identified genes also increase the risk but, taken together, these known mutations account for less than 25 percent of the overall familial risk of breast cancer. High-risk germ-line mutations for colon cancer occur in APC and in mismatch repair genes but, as with breast cancer, known mutations are responsible for only a small proportion (less than 6 percent) of cases. These and similar observations have led to the conclusion, mentioned in chapter 5, that susceptibility to hereditary cancers comes mainly from the combined effects of many genetic events, each one conferring only a small increase in relative risk. The way of getting at this problem is to sequence all the DNA of a large number of people with the disease and compare the results with a similar number of normal sequences. This kind of study requires several thousand individuals in each group to give results that are reliable enough to identify variations (i.e., SNPs) that predispose to specific cancers.

These methods have now identified SNPs that contribute to a wide range of diseases, including most major cancers. These variants are just like other polymorphisms that make human beings different from one another. However, for the disease-associated SNPs the difference is that the base change at the SNP location has the effect of very slightly increasing your chance of getting that disease. The increased risk is very small. For breast cancer SNPs that increase susceptibility have been identified in four different genes but, taken together, they still account for less than 4 percent of the familial risk. Low though the risk is, from a therapy viewpoint it would be nice to be able to counter the effects of these pre-disposing SNPs but, apart from the very rare cases in which they affect the coding sequence of a protein, it is technically difficult even to find out what they do, let alone design a counter-acting drug.

The ease with which individual genomes can now be analyzed also permits the identification of mutations responsible for relatively rare diseases, and the genomes of a number of individuals with a variety of genetic diseases have already been sequenced.

TAILORING THERAPY ON THE BASIS OF WHOLE GENOME SEQUENCING

We've seen that high throughput sequencing has already made important contributions to cancer diagnosis and prognosis and begun to inform the process of drug design. It's also shown that we need to know not just that a gene is mutated but what the precise change is. One important consequence of designing therapy on the basis of a complete mutational picture is that patients can avoid being subjected to treatment regimes that are almost certain to be of no use.

The other point about DNA sequencing is the enormous amount of information it provides. This is both good news and bad news. It hardly needs saying that, in the long run, the more we know, the more comprehensive is our picture of individual cancers and the better placed we are to devise the best treatment package. On the other hand, however, there is the challenge of sifting the most useful information from the vast amount of mutational data. In particular, it's important to differentiate between major cancer genes, that is the "drivers" (e.g., MYC), and the majority that are not essential for cancer development (that is, "passenger" mutations). Nevertheless, the identification of novel cancer genes and mutation signatures for individual tumors will guide efforts to design specific therapies for each.

The most important limitation of all, of course, is the range of drugs available. As we've noted, at the moment there aren't that many specific agents, although the number is increasing, and sequencing is revealing more specific mutations for the pharmaceutical industry to target. If we go back to our Metro map and flag the "stations" for which specific inhibitors already exist, the picture looks quite cheerful in that there are a number of "druggable" targets (several receptors, the MAPK pathway and other pathways, facets of metabolism, and blood vessel function) and, in addition, drugs now exist that can drive cells to suicide. The use of combination therapy can be particularly powerful, especially when distinct pathways are targeted simultaneously. Such ploys have been called "horizontal" inhibition (i.e., of parallel pathways) in contrast to "vertical" inhibition (i.e., of more than one protein in a single pathway).

The main message is that, even with the current contents of the drug cupboard, it's feasible to come up with combinations that target several pathways and that WGS can now tell us which to go for and, equally important, when a particular drug won't work.

One other problem is worth flagging: drugs may block tumor development in some circumstances but have the opposite effect in others, actually enhancing growth or metastasis. This has happened with some anti-vascular agents and it's not yet too clear why. Nevertheless, these problems indicate why drug combinations can be much more effective than single agents, and they also hint at the importance of how drugs are administered. That is, for a given condition there may be advantages to treatment in discrete bursts, as opposed to continuous administration, or the frequent administration of very low doses over prolonged periods, a method called metronomic chemotherapy.

CHIPS TO DETECT TUMORS

One of the most exciting recent developments has been the application of silicon chip technology to the detection of circulating tumor cells (CTCs)—cells that have detached from a primary tumor and entered the bloodstream. The chips resemble a cribbage board except that there are about 80,000 "pegs," the "board" is a glass slide, and the whole thing is about an inch wide and three inches long. The 80,000 tiny columns are etched on the glass surface and coated with an array of antibodies that recognize cell surface proteins. It is enclosed so that a blood sample can be made to flow past the capture columns. Such is the sensitivity with which the antibody coat can pick up passing cells that about 100 CTCs can be captured from a teaspoon (5 ml) of blood. It seems almost incredible that antibodies on chips can pull out CTCs that are present in tiny amounts in circulating blood—as small an amount as one in a billion normal cells. Nevertheless, this advance may offer both the most promising route to early tumor detection, picking out CTCs from the circulation as biomarkers, and of determining responses to chemotherapy. It also provides a bridge between proteomics and genomics because DNA, extracted from the captured cells, can be used for whole-genome sequencing. If this system is able to capture cells from most major types of tumors, it offers a rapid route from early detection through genomic analysis to tailored chemotherapy without the requirement for tumor biopsies.

DRUG SCREENING AND DEVELOPMENT

The upshot of the sequencing revolution is that we will soon have a complete view of the cancer scene in that essentially all mutations will be known. This picture will include patterns that characterize metastases. Complete genomic sequences have already been obtained for a primary tumor and a metastasis that was discovered nine years after the primary had been treated. Given that primary tumors can generally be treated by surgery, the biggest challenge is to find treatments for metastases. This will shift the focus to the problem of developing treatments: how to produce agents that can specifically target dominant oncoproteins and how to reactivate lost tumor suppressors. For the first category another remarkable revolution is under way. It is now possible to generate huge chemical libraries of hundreds of thousands of related compounds. In addition, high-throughput screening systems now mean that the time from making a compound to showing that it does something useful is relatively short.

TUMORS ARE A MESS

At this point we should remind ourselves of a feature of cancer that has recurred in this story, namely that tumors are a hodgepodge. That is, not only is the complete mutational pattern revealed by the whole genome sequence unique to each cancer, but it actually represents an average from a mixed population of cells—a reflection of all the clones that make up a tumor. The implications of this picture are immense because it means that any tumor is made up of groups of genetically distinct cells that are continually evolving—i.e., mutating. Therapeutic strategies are therefore confronting not only mixed targets but moving targets. At first sight, the concept of a moving therapeutic target composed of multiple cancer-propagating cells is somewhat daunting because, intuitively, it would seem easier to have to eliminate just one set of identical cells. However, all may not be lost if the shifting mutational pattern mainly reflects different components in key signal pathways. In other words, we might be able to focus on drug cocktails that target the key pathways, thereby controlling the disease, despite there being different mutational signatures in the clones of the tumor.

ONE FINAL COMPLICATION

As in Haldane's day, surgery remains the first and most effective treatment for many cancers. Furthermore, of course, it remains true that early detection permitting rapid treatment of cancer provides the highest probability of a successful outcome. This in turn suggests that mass screening programs can only be beneficial. Like most aspects of cancer, however, screening methods are not always completely straightforward, and their overall value has been called into question by several recent studies. One of these concluded that in Denmark, decreases in deaths from breast cancer were more likely to be the result of changes in risk factors and improved treatment, rather than of screening. This prompted the *British Medical Journal* to commission an independent look at UK data that also concluded that the benefit of screening was very small. Similar concerns have also been raised over screening for prostate cancer, and we have already noted the shortcomings of the PSA test for this disease. These findings focus on the problem that, although early detection undoubtedly

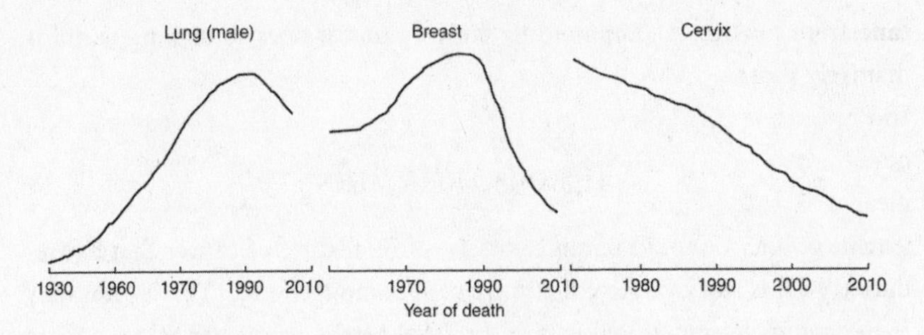

Three ways to get, and get rid of, cancer. *As we saw in chapter 2, smoking causes most lung cancers (left). The decrease in the number of smokers since 1990 has been reflected in a decrease in lung cancer deaths, both in the UK and the US. For breast cancer many things have contributed to the falling death rate since 1990, each having a small effect—a case of every little bit helps, in contrast to the single factor in lung cancer. Cervical cancer represents a third scenario in which the disease arises mainly through viral infection. The continuing downward trend in annual deaths over the last forty years (from 3,000 to about 1,000 in the UK) is largely due to detection by screening at a very early stage when it can be treated—a case of the sooner, the better.*

benefits some, a significant number may be inappropriately diagnosed and treated and that, overall, we need to consider carefully how we fund and carry out screening programs.

ONE FINAL WORD

In charting the history of chemotherapy we have seen that great ingenuity has generated an ever-expanding range of approaches to complement surgery and radiotherapy. There have been some terrific achievements, even though the number of cancers that can be effectively treated at the moment remains small. Hand in hand with these advances, methods for imaging tumors and detecting biomarkers are developing apace. Most dramatically of all, the sequencing revolution has taken us into the era of personalized medicine that is already making an impact on the treatments given to patients and will transform all our lives. One of the anticipated benefits is that diagnosis will become a more precise science. Taken together, the quite staggering scientific advances of the last twenty years hold the ultimate promise that we will be able to detect the threat of cancer in an individual many years before the disease would show any overt signs, merely by the donation of a teaspoon of blood. That, in turn, holds the promise that drugs could displace surgery as the first line of defense, assuming of course that we can come up with enough drugs. The fight against cancer is unending, but the extraordinary events related in this story reveal that in all its long history, the phase we are now entering is the most exciting and promising of all. The alluring vision is, therefore, that although we may never be able to eliminate cancer, the extraordinary triumphs of science would mean its disappearance as a disease. What a wonderful memorial that paradox would be to the men and women who have done battle with the disease over the centuries and to whose efforts this book is dedicated.

GLOSSARY

ADJUVANT THERAPY: A treatment given in addition to the main treatment. For cancer this typically means radiotherapy and chemotherapy after surgery.

ALLELE: One of two (or more) forms of a gene. Genes occupy a specific site in a chromosome. For each pair of chromosomes there are therefore two copies of each gene (i.e., two alleles). The two alleles may be identical (in DNA sequence) or they may differ.

ALPHA PARTICLE: Two protons and two neutrons.

APOPTOSIS: A controlled program by which cells can commit suicide, important in development and also a major protection against cancer.

ATOM: Basic unit of matter: a nucleus surrounded by a cloud of electrons.

ATP: Adenosine triphosphate: a nucleotide made up of a sugar (ribose) to which is attached the base adenine (A) and three phosphate groups. It is the major way in which energy is transferred within cells and is also one of the four units incorporated in DNA.

BENIGN TUMOR: Such tumors may arise in any tissue and cause local damage by pressure or obstruction but do not spread to other sites or invade adjoining tissues.

BETA PARTICLE: Electrons emitted by nuclei.

CACHEXIA: Loss of body weight in someone not trying to lose weight. Also called wasting syndrome, it often occurs in cancer.

CARCINOGEN: Any substance that can contribute to the development of cancer. Carcinogens include chemicals, radiation, and radionuclides that can damage DNA directly as well as substances that can promote cancer by perturbing metabolism.

CELL: The most basic unit of organisms that can survive independently. Animal cells are bounded by a flexible sac called the plasma membrane. Membranes also form separate compartments within cells, e.g., the nucleus and the mitochondrion.

CHROMOSOME: The nucleus of human cells contains two meters of DNA divided into 23 pairs of chromosomes. Chromosomes are numbered 1 to 22 in decreasing order of length. In addition there are two sex chromosomes, X and Y. Chromosomal DNA folds up by wrapping around proteins.

COMPUTED TOMOGRAPHY: A method that gives a three-dimensional picture of tissues in the body from a large series of two-dimensional X-ray images.

DNA: Deoxyribonucleic acid: a nucleic acid that contains the genetic instructions that make all living cells. It is a polymer of simple units called nucleotides. In the nucleus it forms a double helix of two intertwined strands. Each unit (nucleotide) is made of a base, a sugar, and a phosphate group. There are four bases: adenine (A), cytosine (C), guanine (G), and thymidine (T), each made of carbon, nitrogen, oxygen, and hydrogen atoms. The backbone of DNA is formed by a link between the phosphate and sugar in adjacent units. The sugar units in DNA (deoxyribose) have one less oxygen atom than those in RNA (ribose). The sequence of bases (A, C, T, and G) carries the genetic information.

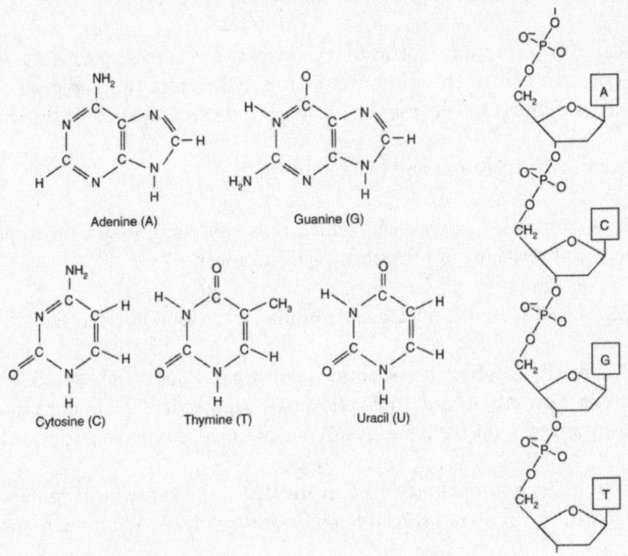

Adenine (A) Guanine (G)

Cytosine (C) Thymine (T) Uracil (U)

Nucleic Acid Bases and DNA

DRIVER MUTATION: Genetic mutation that promotes a stage or stages of tumor development.

ENDOTHELIAL CELLS: Cells that line the inner surfaces of all blood vessels; the whole lining is called the endothelium.

EPITHELIAL CELLS: Cells that line the inner and outer surfaces of the body in continuous sheets, sometimes called epithelial membranes or epithelia. The epithelium is one of the four basic types of animal tissue, along with connective tissue, muscle tissue, and nervous tissue. There are four main classes of epithelium: (1) simple squamous, (2) simple cuboidal, (3) simple columnar, (4) pseudostratified. Squamous cells are flat cells that are part of the epithelium as either a single layer (simple squamous epithelium) or multiple layers (stratified squamous epithelium). Pseudostratified cells are a form of column-shaped cells that may also be ciliated, i.e., have fine hair-like extensions. Ciliated epithelium occurs in several places including the nose.

Eukaryotic cell: A cell that contains within its outer membrane smaller, specialized compartments that are themselves bounded by membranes, e.g., the nucleus or mitochondria. This distinguishes them from prokaryotes that do not have a nucleus. Bacteria are single-cell prokaryotes.

Fatty acids: Fatty acids are unbranched chains of carbon atoms with a methyl group (CH_3–) at one end and a carboxyl group at the other (–COOH). Thus they are a molecule of acetic acid with additional CH_2 groups inserted between the CH_3– and the –COOH ends. The methyl (CH_3– end) carbon of a fatty acid is sometimes called the omega (ω carbon: an omega–3 fatty acid has its first double bond on the third carbon counting from the ω (methyl) end. Humans cannot make ω–3 fatty acids so they are an important part of our diet.

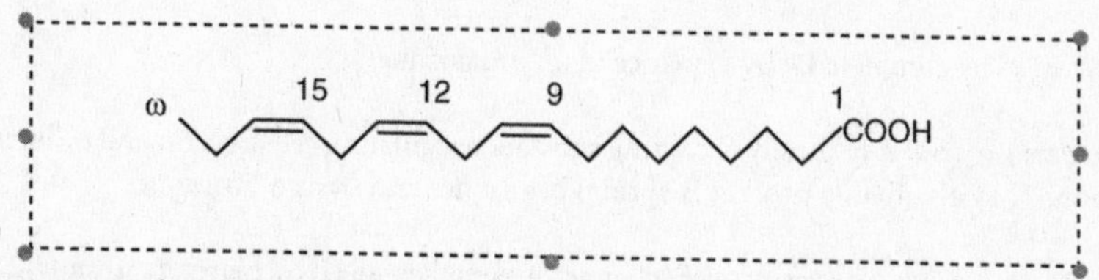

An omega–3 fatty acid (alpha-linolenic acid): *Each kink represents a carbon atom with attached hydrogen atoms.*

In membranes the most common fatty acids have 18 carbon atoms: they are usually unsaturated, that is, they contain at least one double bond (that's two carbon atoms joined by two rather than one covalent bond). One double bond makes them monounsaturated; if they contain more than one, they are polyunsaturated. Double bonds may be either *cis* or *trans* (hydrogen atoms on the same side or on opposite sides of the double bond, respectively). Naturally occurring fatty acids are rich in the *cis* forms. Unsaturated fats are considered healthier than saturated fats because they lower total cholesterol and LDL cholesterol levels in the blood but *trans* unsaturated fats are to be avoided because the form of the double bond permits a linear structure that may promote plaque formation.

Cis double bond H H C = C H C = C H *Trans* double bond

Gene: A unique sequence in DNA that encodes a protein or RNA. Genes are the basic unit of inherited information.

GENETIC INSTABILITY: A cellular state in which mutations occur continuously at an abnormally high rate. Instability may be manifested from the single nucleotide level to gains or losses of whole chromosomes.

GENETIC RECOMBINATION: The breakage and rejoining of parental chromosomes (M and F) generates chromosomes C1 and C2 that carry DNA from both parents.

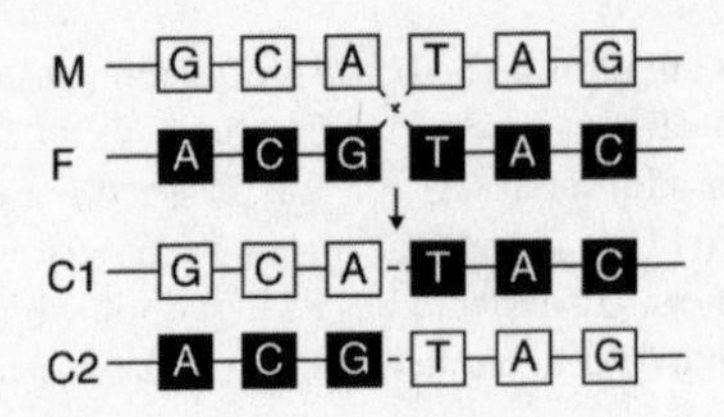

GENOME: The complete DNA sequence of an organism.

GROWTH FACTOR: A naturally occurring substance regulating cell division and differentiation. They are mainly protein (peptide) hormones and steroid hormones.

HOMOLOGOUS RECOMBINATION: Exchange of stretches of DNA between strands that are similar or identical in sequence (i.e., homologous chromosomes)—called crossing over. It requires the strands to be broken in a chromosome from each parent to allow segments of DNA to swap chromosomes. The exchanged segments of DNA are then joined to their new chromosomes to produce novel combinations of genes.

HYPOXIA: Condition in which all (generalized hypoxia) or parts (tissue hypoxia) of the body experience reduced oxygen supply. Complete oxygen deprivation is referred to as anoxia.

IMMUNE SYSTEM: The organs, tissues, cells, and proteins that enable the body to resist infection. When infection occurs, there are two main parts to the defensive response: the innate immune system and the adaptive immune system. The first is an immediate, nonspecific response. The adaptive immune response is activated by the innate system and produces improved recognition of the infectious agent that is retained after the cause has been eliminated—an effect called immunological memory. The immune system forms an important defense against cancer because tumor cells express abnormal proteins and thus appear "foreign," that is, the body responds as it would to an infection.

ISOTOPE: A variant of a chemical element due to different numbers of neutrons. The number of protons is unchanged. For example, there are three isotopes of carbon (–12, –13, and –14): each has six protons, so their atomic number is 6, but the isotopes have 6, 7, and 8 neutrons.

KINASE: Enzymes that transfer phosphate groups from donor molecules (particularly ATP) to specific substrates. The process is called phosphorylation, and kinases may therefore also be called phosphotransferases.

LYMPHOCYTE: A type of white blood cell that functions in the immune response.

LYMPHOID CELLS: One of the major classes of cells derived from bone marrow stem cells.

LYMPHOMA: Malignant proliferation of lymphoid cells. Major divisions of the disease are Hodgkin's disease (Hodgkin's lymphoma) and non-Hodgkin's lymphoma.

MALIGNANT TUMOR: Uncontrolled cell growth characterized by invasion through the basement membrane into surrounding tissue and a propensity to spread (metastasize) by blood or lymphatic routes to other sites.

MEIOSIS: The conversion of one diploid cell (containing two copies of each chromosome—one from mother, one from father) to four haploid cells (containing one copy of each chromosome). These are sperm and egg cells (gametes). Each cell has *either* maternal *or* paternal chromosomes, apart from small segments that have been exchanged by homologous recombination.

MESSENGER RNA: A copy of the sequence of DNA that is used to produce the corresponding protein.

METASTASIS: Spread of malignant cells from the site of origin to other sites.

MITOGEN: A chemical that acts on cells to promote growth and cell division.

MOLECULE: A group of atoms held together by strong (covalent) chemical bonds.

MONOCLONAL ANTIBODY: An antibody is a protein made by the immune system that neutralizes bugs. Monoclonal antibodies are made in the laboratory from identical immune cells cloned from a unique parent cell.

MUTATION: Any change in the sequence of DNA. Mutations range from a change of one base to another to complete loss of a chromosome or amplification to produce multiple copies of a chromosome.

NEOPLASM: New or abnormal cell growth, e.g., a tumor.

ONCOGENE: A mutated version of a normal gene (called a proto-oncogene). The abnormal protein produced perturbs normal cell control and helps promote cancer.

PARASITE: Any living thing that lives in or on another living organism.

PET: Positron emission tomography; a method that uses radioactive tracers to produce a three-dimensional picture of tissues in the body, particularly used to detect tumors.

PHOTON: Particles without mass that comprise electromagnetic radiation, a fluctuating electric, and magnetic field that travels at the speed of light. Their energy is proportional to their frequency: thus radio waves have photons with low energies and are harmless to living matter whereas X-rays and gamma-rays are photons with enough energy to damage molecules.

PROTEIN: A large molecule formed by joining simple units (amino acids) together. Apart from water, proteins are the major components of cells.

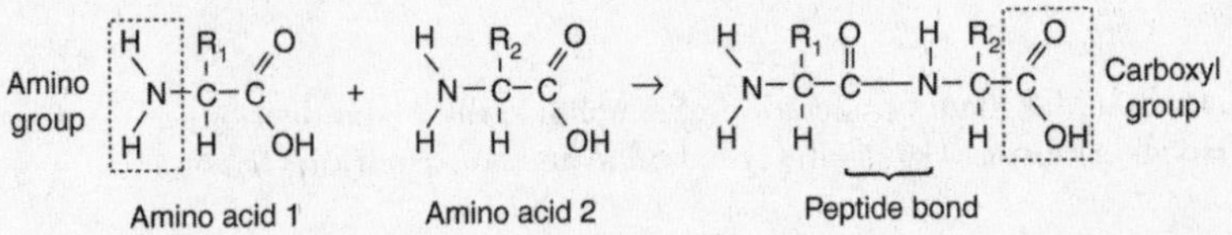

Amino group
Carboxyl group
Amino acid 1
Amino acid 2
Peptide bond

Two amino acids joining via a peptide bond. *The twenty amino acids differ in the groups shown as* R_1 *and* R_2*. The smallest R group (a hydrogen atom) gives the amino acid glycine. The largest is in tyrosine and contains a phenol ring.*

RADICALS: Radicals (usually called free radicals) are atoms, molecules, or ions that have unpaired electrons (an electron in an orbital, rather than an electron pair). Reactive oxygen species are free radicals.

RETROVIRUSES: Family of viruses with RNA (rather than DNA) genomes.

SENESCENCE: Biological aging in which cells stop dividing: for this reason it is also a protection against cancer.

STEM CELL: Cells present in all multicellular organisms that have two central properties: (i) they can go through numerous cycles of division to produce more stem cells, and (ii) they possess potency—the capacity to form more specialized cell types.

TELOMERASE: An enzyme that adds repeated sequences of DNA (called telomeres) to the ends of chromosomes. Its activity is lost in adult cells as part of the aging process, but it is active in most tumor cells, enabling them to grow indefinitely.

TRANSCRIPTION FACTORS: Proteins that bind to specific DNA sequences to regulate gene expression, the transcription of DNA into RNA.

TUMOR SUPPRESSOR: A protein whose loss or inactivation may promote cancer.

GENE NAMES

AT A VERY EARLY STAGE IN THE ONCOGENE STORY (1980), WITH A FORESIGHT rare in the annals of science, the leading protagonists got together to lay down a set of rules governing how cancer genes should be named. Their recommendation was a three-letter code in italics for each gene name with the name being such that it could not be rendered inappropriate or misleading by any subsequent discoveries about its function. Thus *sarc* became "*SRC*," the abbreviation reflecting the fact that it had been identified as a cause of sarcomas. The viral form of the gene became v-*src*.

Most human genes stick to this rule, although there are some four-letter names and close family members are sometimes distinguished by suffixed numbers.

Strictly, gene names are written in italicized capitals, but the protein that they encode is non-italicized; *EGFR* (gene)/EGFR (protein), as an example. They are pronounced phonetically when possible (SRC is *sarc*, MYC *mick*, ABL *able*).

In this book, for simplicity's sake, we've broken the rules and just used non-italicized capitals for both proteins and genes.

SOURCES AND RESOURCES

CANCER STATISTICS

Global

Union for International Cancer Control (UICC): http://globocan.iarc.fr/
World Health Organization (WHO): http://www.who.int/cancer/en/
WHO, Global InfoBase, "Cancer Country Profiles": http://www.who.int/infobase/report.aspx?rid=126
Center for Communications, Health and the Environment: http://www.ceche.org/
CancerBACUP: http://www.jamkit.com/Clients/Nonprofits/CancerBACUP

UK and USA

National Cancer Institute: http://www.cancer.gov/
American Cancer Society: http://www.cancer.org/
Cancer Research UK: http://info.cancerresearchuk.org/cancerstats/

Europe

European Cancer Observatory: http://eu-cancer.iarc.fr/

Latin America

The Pan American Health Organization (PAHO): http://new.paho.org/

ELECTRICITY SUPPLIES

http://en.wikipedia.org/wiki/Mains_electricity_by_country

SKIN CANCER

American Academy of Dermatology: http://www.aad.org/skin-conditions/skin-cancer-detection/about-skin-self-exams

BREAST TISSUE DENSITY MAMMOGRAMS

http://www.mayoclinic.com/health/medical/IM03415

CHEST X-RAYS

http://www.chestx-ray.com/genpublic/genpubl.html

http://www.pennmedicine.org/encyclopedia/em_PrintArticle.aspx?gcid=003804&ptid=1

FLUORESCENT JELLYFISH

http://www.flickr.com/photos/gregory-moine/3100736911/

FLUORESCENT ZEBRA FISH

http://www.tropicalfishandaquariums.com/Carp/Glofish.asp

FLUORESCENT WORM

http://news.preview.nationalgeographic.com/news/2009/05/photogalleries/glowing-animal-pictures/photo9.html

WORMATLAS

http://www.wormatlas.org/

MANUSCRIPT COPYING AND GENES

http://www.plosbiology.org/article/info%3Adoi%2F10.1371%2Fjournal.phio.1001069

HUMAN GENOME PROJECT INFORMATION

http://www.ornl.gov/sci/techresources/Human_Genome/home.shtml

BROWSE YOUR GENOME

http://www.ncbi.nlm.nih.gov/genome/guide/human/

TUMOR GRADING AND STAGING

A data base for the TNM system is maintained by the International Union Against Cancer (UICC: http://www.uicc.org/) as part of its remit to integrate the efforts of cancer organizations world-wide.

SURGERY, SYMPTOMS, AND CHEMOTHERAPY

Information on surgery: http://www.mayoclinic.com/health/cancer-surgery/CA00033.
Cancers: symptoms, treatment and outlook: CRUK: http://www.cancerhelp.org.uk/
Chemotherapy: Chemocare.com: http://www.chemocare.com/bio/list_by_acronym.asp
eTUMOUR study: http://www.etumour.net/

CANCER GENES

The Cancer Genome Atlas (TCGA): http://cancergenome.nih.gov/

The Cancer Gene Census catalogue of genes for which mutations have been causally implicated in cancer: http://www.sanger.ac.uk/genetics/CGP/Census/

VIDEO LINKS

DNA makes RNA makes Protein
http://www.youtube.com/watch?v=D3fOXt4MrOM

Cell division
http://www.youtube.com/watch?v=rgLJrvoX_qo&feature=related

The Cell Cycle
http://www.youtube.com/watch?v=lf9rcqifx34&feature=related

Proto-oncogene to Oncogene
http://www.youtube.com/watch?v=2wIVwZksIt4&feature=related

How Cancer Develops
http://www.youtube.com/watch?v=A1Fkdt–2veM&feature=related

Tumor growth
http://www.dnalc.org/resources/3d/

Blood vessel growth (Angiogenesis)
http://www.youtube.com/watch?v=qghs7ZvetbA&feature=related
http://www.dnatube.com/video/1171/Full-VEGF—Angiogenesis-Video

Metastasis
http://www.youtube.com/watch?v=rrMq8uA_6iA

Metastasis in Motion
http://www.niehs.nih.gov/news/video/scivid/mmp1.cfm
http://www.niehs.nih.gov/news/video/scivid/mmp2.cfm

Pathways to Cancer
http://www.dnalc.org/resources/3d/pathways.html

MAPK pathways
http://www.biocreations.com/animations/MAP_Kinase.swf

BOOKS

Each of these provides additional background to the cancer story. They are all enjoyable and readily understandable by non-scientist readers.

Bishop, Michael. *How to Win the Nobel Prize: An Unexpected Life in Science.* Cambridge, Mass.: Harvard University Press, 2003.
Goldacre, Ben. *Bad Science.* London: Fourth Estate, 2008.
Jones, Steve. *The Language of the Genes.* London: Flamingo, 1993.

Judson, Horace. *The Eighth Day of Creation.* 25th anniversary ed. Plainview, N.Y.: Cold Spring Harbor Laboratory Press, 2004.
Schoenfeld, Robert. *The Chemist's English.* Weinheim, Germany: VCH, 1990.
Shubin, Neil. *Your Inner Fish: A Journey into the 3.5-Billion-Year History of the Human Body.* New York: Pantheon Books, 2008.
Skloot, Rebecca. *The Immortal Life of Henrietta Lacks.* London: Macmillan, 2010.
Sulston, John, and Georgina Ferry. *The Common Thread: A Story of Science, Politics, Ethics, and the Human Genome.* London: Corgi, 2003.
Watson, James. *The Double Helix.* London: Phoenix, 2010.
Wilkins, Maurice. *Maurice Wilkins: The Third Man of the Double Helix: An Autobiography.* Oxford and New York: Oxford University Press, 2003.
Witherly, Jeffre L., Galen P. Perry, and Darryl L. Leja. *An A to Z of DNA Science: What Scientists Mean When They Talk about Genes and Genomes.* Cold Spring Harbor, N.Y.: Cold Spring Harbor Laboratory Press, 2001.

PAPERS

These are the references for four papers on DNA, published in the science journal *Nature,* that have had a major impact on the story of cancer.

DNA structure

Watson J. D., and F. H. C. Crick. "Genetical Implications of the Structure of Deoxyribonucleic Acid," *Nature* 171 (1953): 964–967.
http://libsta28.lib.cam.ac.uk:2087/nature/journal/v171/n4361/pdf/171964b0.pdf

DNA Sequencing

Sanger, F., A. R. Coulson, T. Friedman, G. M. Air, B. G. Barrell, N. L. Brown, J. C. Fiddes, C. A. Hutchison, P. M. Slocombe, and M. Smith, "Nucleotide Sequence of Bacteriophage phi X174 DNA," *Nature* 265 (1977): 687–695.
http://libsta28.lib.cam.ac.uk:2087/nature/journal/v265/n5596/pdf/265687a0.pdf

The Human Genome Sequence

International Human Genome Sequencing Consortium, "Initial Sequencing and Analysis of the Human Genome," *Nature* 409 (2001): 860–921.
http://www.nature.com/nature/journal/v409/n6822/full/409860a0.html
International Human Genome Sequencing Consortium, "Finishing the Euchromatic Sequence of the Human Genome," *Nature* 431 (2004): 931–945.
http://libsta28.lib.cam.ac.uk:2122/nature/journal/v431/n7011/pdf/nature03001.pdf

DRUG INFORMATION

The National Cancer Institute provides a comprehensive drug data base including information on clinical trials (http://www.cancer.gov/DRUGDICTIONARY/). Lymphomation.org (http://www.lymphomation.org/chemo-index.htm) also provides an extensive listing with links to other sites. Numerous other sites are listed on the DrugInfo site (http://www.drugsrnikolov.com/drugdatabases.html).

"CANCER IS A FUNNY THING"

This poem by J.B.S. Haldane was published in *The New Statesman,* London on February 21, 1964 (p. 298) and is reproduced by kind permission of *The New Statesman.*

ODES TO CANCER

1964

I wish I had the voice of Homer
To sing of rectal carcinoma,
Which kills a lot more chaps, in fact,
Than were bumped off when Troy was sacked.
Yet, thanks to modern surgeons' skills,
It can be killed before it kills
Upon a scientific basis
In nineteen out of twenty cases.
I noticed I was passing blood
(Only a few drops, not a flood).
So pausing on my homeward way
From Tallahassee to Bombay
I asked a doctor, now my friend,
To peer into my hinder end,
To prove or disprove the rumour
That I had a malignant tumour.
They pumped in $BaSO_4$
Till I could really stand no more,
And, when sufficient had been pressed in,
They photographed my large intestine.
In order to decide the issue
They next scraped out some bits of tissue.
(Before they did so, some good pal
Had knocked me out with pentothal,
Whose action is extremely quick,
And does not leave me feeling sick.)
The microscope returned the answer
That I had certainly got cancer.
So I was wheeled into the theatre
Where holes were made to make me better.
One set is in my perineum
Where I can feel, but can't yet see 'em.
Another made me like a kipper
Or female prey of Jack the Ripper.

2011

Long gone are the days of Homer
But not so those of carcinoma,
Of sarcoma and leukemia
And other cancers familia.

But nowadays we meet pre-school
That great and wondrous Molecule.
We know now from the knee of Mater
That DNA's the great creator.
Young Jim Watson and F. Crick
Showed that's what does the trick,
Though clever Sydney Brenner said:
"It needs, lest info gets misled,
A messenger—a go-between
To what Max calls the life machine."
So there it was: The Dogma Central
The thing that puts us on one level.
All the bugs and birds and bees,
And of course the plants and trees,
All the proteins that make you and me
Everything starts with T, G, A, and C.
But bases four—is that enough?
More than—it's quite degenerate stuff.
Just as weird—I know it shocks
That only twenty building blocks
Can make an infinite protein stash
With which Ma Nature's had a bash
At species of the earth galore
And those that snuffed it heretofore.
But DNA makes cancer too
Time enough—it'll happen to you.
"No worries sport" as some would say,
These days it's "omics" all the way.
So heed the words of JBS
Who years ago, though in distress,

Through this incision, I don't doubt,
The neoplasm was taken out,
Along with colon, and lymph nodes
Where cancer cells might find abodes.
A third much smaller hole is meant
To function as a ventral vent:
So now I am like two-faced Janus
The only god who sees his anus.*
*(*In India there are several more*
With extra faces, up to four,
But both in Brahma and in Shiva
I own myself an unbeliever.)
I'll swear, without the risk of perjury,
It was a snappy bit of surgery.
My rectum is a serious loss to me,
But I've a very neat colostomy,
And hope, as soon as I am able,
To make it keep a fixed time-table.
So do not wait for aches and pains
To have a surgeon mend your drains;
If he says 'cancer' you're a dunce
Unless you have it out at once,
For if you wait it's sure to swell,
And may have progeny as well.
My final word, before I'm done,
Is 'Cancer can be rather fun.'
Thanks to the nurses and Nye Bevan
The NHS is quite like heaven
Provided one confronts the tumour
With a sufficient sense of humour.
I know that cancer often kills,
But so do cars and sleeping pills
And it can hurt one till one sweats,
So can bad teeth and unpaid debts.
A spot of laughter, I am sure,
Often accelerates one's cure;
So let us patients do our bit
To help the surgeons make us fit.

J.B.S. Haldane

Gave this advice on what to do
When something odd happens to you:
"Take blood and bumps to your physician
Whose only aim is your remission."
But now it's different 'coz you'll see
Not sliced but sequenced you will be.
You and your tumor—what a pair!
In times of yore you'd not a prayer
But how things change, you must agree,
It's genomics—twenty-first century!
Personal medicine—just what it says
The keyboard of genes that cancer plays.
With all that gen those wondrous medics
No longer battle with their ethics.
They know which are delinquent genes
And how to zap 'em to smithereens!
Down in one and tickety-boo!
Farewell tumor, it's toodle-oo.
Well maybe not yet—give us a chance!
Sequence data's an avalanche!!
Lots of new targets—wow, what drama!
So put all your faith in Big Pharma.
And if we're right put your cash in too
The future's chemo for you—and you!!

DEDICATION

This book is dedicated to Marie-Adèle Rajandream, a gifted scientist who was a member of the Laboratory of Molecular Biology before moving to the Sanger Centre, Cambridge, in 1993. She was diagnosed with cancer in the summer of 2007 and died on July 12, 2008, working from her bed with her colleagues until the previous day.

ACKNOWLEDGMENTS

ANY BOOK ON THE SCIENCE OF CANCER OWES AN INESTIMABLE DEBT to two groups of people. First are the nameless legions who through the years have fought their personal battle with the disease and while doing so have freely contributed themselves to scientific research. Their selflessness has made an incalculable contribution to what we know, increasingly so as we move into the era of mass screening and personalized medicine. Second are all the scientists and physicians whose labors have brought us to where we are and whose efforts are enshrined in their published work. I have known a tiny proportion of these personally and I have drawn on countless hours of discussion with them as well as on the literature. The number who deserve acknowledgement thus runs into thousands, represented here, I'm afraid, by only the small number mentioned by name. I owe a special debt to three colleagues with whom I've worked for many years: Jim Metcalfe, who taught me how to think about science; Gerry Smith, a wonderfully gifted organic chemist; and Heide Kirschenlohr, the most unfailingly rigorous experimentalist I know. In addition I am profoundly grateful to more generations of students than I care to count for their stimulating input, in particular for their uncanny knack of asking incisive questions that I can't answer.

This book would not have seen the light of day without the amazing enthusiasm and organizational powers of Luba Ostashevsky of Palgrave Macmillan. It has also been a privilege to collaborate with the authoress Lynn Vannucci in the final stages of editing. Thomas Shafee, a graduate student in the Department of Biochemistry, Cambridge, drew the sketches with great skill and ingenuity.

The following have made greatly valued contributions either through critically reading the manuscript, providing encouragement, giving expert advice or permitting their photographs to be reproduced: Bart Barrell, Tom Booth (Cambridge Research Institute), Peter Börnert (Philips Technologie GmbH, Hamburg), Sir Michael Brady (Department of Oncology, University of Oxford), Peter Britton (Addenbrooke's Hospital), John Buscombe (Division of Nuclear Medicine, Addenbrooke's Hospital, Cambridge), Jean Chothia (Faculty of English, University of Cambridge), Alan Coulson, Ferdia Gallagher (Department of Radiology, Addenbrooke's Hospital, Cambridge and Cancer Research UK Cambridge Research Institute), John Griffiths (Cambridge Research Institute), Brian Huntly (Cambridge Institute for Medical Research), Becky Lloyd, Sir Alex Markham (University of Leeds), Rahmi Oklu (Massachusetts General Hospital), Ashwin Rattan (Searching Finance), Fred Sanger, Richard Sever (Cold Spring Harbor Laboratory Press), Sir John Sulston (Institute for Science, Ethics and Innovation, University of Manchester), Robert Tasker (Harvard Medical School), Rupert Thompson (Faculty of Classics, University of Cambridge), Sir John Walker (MRC Mitochondrial Biology Unit, Cambridge), Robert Whitaker (Selwyn College), Richard White (Dana Farber Cancer Institute, Children's Hospital Boston), Roger Wilkins, Rick Wilson (Washington University, St. Louis), and from the Department of Biochemistry, University of Cambridge, David Ellar, Gerard Evan, Richard Farndale, Chris Green, Jules Griffin, Chris Howe, and Tom Mayle.

My sons Robert (Select English) and Richard (Cardiff University School of Medicine) have been invaluable proofreaders and critics but nevertheless remain two of my best friends. My other best friend is my wife who quite simply makes everything worthwhile and even finds the energy in her *alter ego* as Jane Rogers to be my tutor in genomics.

INDEX